Irrigation Handbook

Irrigation Handbook

Natalie Ramsey

MURPHY & MOORE
www.murphy-moorepublishing.com

Published by Murphy & Moore Publishing,
1 Rockefeller Plaza,
New York City, NY 10020, USA
www.murphy-moorepublishing.com

Irrigation Handbook
Natalie Ramsey

International Standard Book Number: 978-1-63987-338-8 (Hardback)

Cataloging-in-Publication Data

Irrigation handbook / Natalie Ramsey.
p. cm.
Includes bibliographical references and index.
ISBN 978-1-63987-338-8
1. Irrigation. 2. Agriculture. 3. Water in agriculture.
I. Ramsey, Natalie.
S613 .I77 2022
631.7--dc23

TABLE OF CONTENTS

PREFACE

The artificial techniques which are employed to apply controlled amounts of water with the objective of assisting agricultural activities fall under irrigation. The discipline finds a wide variety of applications such as maintenance of landscapes, revegetation of disturbed soils in dry areas, frost protection, suppression of weed growth, and prevention of soil consolidation. According to the method of water supply and the amount of water supplied, irrigation techniques can be classified into surface irrigation, micro irrigation, sprinkler irrigation and sub irrigation. Sprinkler irrigation can be further divided into methods using central pivot, lateral move, lawn sprinklers and hose-end sprinklers. Some of the different water sources which are used for irrigation systems are springs, wells, rivers, lakes, drainage water and treated wastewater. This book elucidates the concepts and innovative models around prospective developments with respect to irrigation. It aims to shed light on some of the unexplored aspects of irrigation. This book is appropriate for students seeking detailed information in this area as well as for experts.

A foreword of all Chapters of the book is provided below:

Chapter 1 - Irrigation is an artificial process of supplying water to land or crops to assist in the production of crops. Irrigation management is the decision process and action of timing and regulating irrigation water applications so that the water requirement of the crops is met without the waste of water. This chapter serves as an introduction to the various aspects of irrigation; **Chapter 2 -** Soil is a complex mixture of minerals, organic matter, air, water and organisms that support life. Soil moisture is the content of water in the soil. Water present in the soil is classified into hygroscopic water, capillary water and gravitational water. The topics elaborated in this chapter will help in gaining a better perspective about the soil-water relations; **Chapter 3 -** The potential energy of water in a system compared to pure water under same temperature and pressure conditions is known as water potential. Osmosis is the movement of solvent molecules from a region of higher concentration to the region of lower concentration through a semi-permeable membrane. The chapter closely examines the key concepts of plant and water relations to provide an extensive understanding of the subject; **Chapter 4 -** There are several methods of irrigation like surface irrigation, micro-irrigation, drip irrigation, sprinkler irrigation, subsurface textile irrigation, center pivot irrigation, garden hose, subirrigation etc. Surface irrigation is the oldest form of irrigation. In this method, water moves across the surface of fields in order to infiltrate into the soil. Various techniques and methods of irrigation have been thoroughly discussed in this chapter; **Chapter 5 -** Water resources are the natural resources of water that are potentially useful to humans. The water present beneath Earth's surface in the spaces of soil is known as groundwater. An aquifer is the saturated zone beneath the water table. This chapter has been carefully written to provide an easy understanding of the various water resources for irrigation.

I would like to thank the entire editorial team who made sincere efforts for this book and my family who supported me in my efforts of working on this book. I take this opportunity to thank all those who have been a guiding force throughout my life.

Natalie Ramsey

Chapter 1

Irrigation: An Introduction

Irrigation is an artificial process of supplying water to land or crops to assist in the production of crops. Irrigation management is the decision process and action of timing and regulating irrigation water applications so that the water requirement of the crops is met without the waste of water. This chapter serves as an introduction to the various aspects of irrigation.

Irrigation is the process of applying water to the crops artificially to fulfil their water requirements. Nutrients may also be provided to the crops through irrigation. The various sources of water for irrigation are wells, ponds, lakes, canals, tube-wells and even dams. Irrigation offers moisture required for growth and development, germination and other related functions.

The frequency, rate, amount and time of irrigation are different for different crops and also vary according to the types of soil and seasons. For example, summer crops require a higher amount of water as compared to winter crops.

Types of Irrigation

There are different types of irrigation practiced for improving crop yield. These types of irrigation systems are practiced based on the different types of soils, climates, crops and resources. The main types of irrigation followed by farmers include following methods:

- Surface Irrigation: In this system, no irrigation pump is involved. Here, water is distributed across the land by gravity.
- Localized Irrigation: In this system, water is applied to each plant through a network of pipes under low pressure.
- Sprinkler Irrigation: Water is distributed from a central location by overhead high-pressure sprinklers or from sprinklers from the moving platform.
- Drip Irrigation: In this type, drops of water are delivered near the roots of the plants. This type of irrigation is rarely used as it requires more maintenance.
- Centre Pivot Irrigation: In this, the water is distributed by a sprinkler system moving in a circular pattern.
- Sub Irrigation: Water is distributed through a system of pumping stations gates, ditches and canals by raising the water table.
- Manual Irrigation: This is a labor intensive and time-consuming system of irrigation. Here, the water is distributed through watering cans by manual labor.

Methods of Irrigation

Irrigation can be carried out by two different methods:

- Traditional Methods
- Modern Methods

Traditional Methods of Irrigation

In this method, irrigation is done manually. Here, a farmer pulls out water from wells or canals by himself or using cattle and carries to farming fields. This method can vary in different regions. The main advantage of this method is that it is cheap. But its efficiency is poor because of the uneven distribution of water. Also, the chances of water loss are very high. Some examples of the traditional system are pulley system, lever system, chain pump. Among these, the pump system is the most common and used widely.

Modern Methods of Irrigation

The modern method compensates the disadvantages of traditional methods and thus helps in the proper way of water usage.

The modern method involves two systems:

- Sprinkler system: A sprinkler system, as its name suggests, sprinkles water over the crop and helps in an even distribution of water. This method is much advisable in areas facing water scarcity. Here a pump is connected to pipes which generate pressure and water is sprinkled through nozzles of pipes.
- Drip system: In the drip system, water supply is done drop by drop exactly at roots using a hose or pipe. This method can also be used in regions where water availability is less.

Importance of Irrigation

- Compensating for When There is no Rainfall: Irrigation kicks in when there is insufficient rainfall or when there is uncertainty as to when the rain will fall. Without the rain or irrigation as the alternative, the crops are adversely affected, which can lead to a food shortage or crop/plant failure.
- Increasing the Amount of Cultivatable or Agriculturally Productive Land: Some areas of the world are dry, naturally. Irrigation has been responsible for turning such lands into cultivatable lands. Today, about 18% of the world's cropland is now irrigated. Irrigation is also responsible for bringing the most out of fallow lands, which are traditionally left idle after harvest, before the next cultivation season.
- Helps Meet Food Demands: The expansion of irrigated land has made desert ecosystems, such as Jordan and Israel, adopt farming to boost food demands without necessarily depending on rain. Such areas use groundwater from wells and aquifers and in doing so, it enables them to meet the world's demand for food by producing food crops such as cereals, potatoes, and vegetables.

- Furthermore, countries and regions that practice top-level and large-scale irrigation are best known for exporting food, which brings an economic advantage to such a region or country.
- Increased Productivity: Irrigation, by default and in many instances, is employed when rainfall is insufficient. However, it can be applied at any time, even if there is sufficient rainfall to boost crop productivity.
- Enables Multiple Cropping: In most tropical countries, the rainy seasons are specific, making it impossible to grow multiple crops at the same time. Moreover, their water requirements are different, and over-irrigation spoils the crop production. Optimum irrigation makes it possible to grow more than one crop at a time in most areas of a country.
- It is Effective: When crops or plants are watered irregularly and with just a mere hosepipe, it is not consistent and they simply get just one pass. This means hard-to-reach areas and those that do not experience steady rainfall may not get sufficient water. A well-designed irrigation system delivers even to hard-to-reach areas of the land by the use of efficient irrigation systems such as drip irrigation.
- Defeating and Efficient Nature: We cannot control when it will rain. If it rains during the day, and it gets a bit warm, evaporation rates will be high; meaning the crops and the soil will not consume enough water.
- Irrigation systems can be set in such a way that they sprinkle water early in the morning or at night when evaporation levels are low. This, not only saves on the water used, but also maximizes the moisture required by the crops, flowers, or even the soil itself.
- Contribution to Economic Growth: Irrigation ensures food production continues, regardless of the season or climatic condition. This means there are continuous income and employment, thereby reducing poverty. The substantial increase in income achieved through irrigation means the economy can continue to thrive. This is also achieved by the export of food to other regions or countries.

Advantages of Irrigation

Followings are the major advantages of irrigation:

- Irrigation helps to maintain the condensation of the loam.
- It provides mineral as well as other nutrition by the assimilation from the mold.
- Irrigation is one of the most feasible ways to grow cash crops like sugarcane, tobacco, etc.
- As people are dependent on irrigation channels, there is now an extensive irrigation system which people are using as a mean of their communication.
- It helps to preserve the structure of the topsoil.
- By eliminating the deficiency of water, Irrigation manages to give fast manufacturing.
- The more production rate intensifies the more living standards of the farmer increases.

- Canal and irrigation water seeps through the soil and elevates the water table. This is desirable in desiccated and semiarid zones.
- At many places, the irrigation serves as the only water source.
- Irrigation helps to make the cash flow of a country profound with the massive production of crops.
- It helps the landowners by the monetary increment of fertile lands.
- People in rural areas use irrigated water for household chores like- dishwashing, bathing.
- Irrigation helps to improve the yielding.
- The irrigation waterways can be utilized to produce hydroelectric power.
- As people are more prone to irrigation now, colossal irrigation channel construction is assisting the people to be employed known as relief workers.
- It helps to prevent the disease of weeds.
- It acts as a shield against famine.
- Prolonged irrigation is used for navigational impetus sometimes.
- The banks of the irrigation channels can be wielded to prevent deforestation and improve contexture status.

Disadvantages of Irrigation

Followings are the major disadvantages of irrigation:

- Irrigation sometimes creates unrestrained excretion and percolation of water throughout the whole passage.
- The overflowing irrigated water gives insects a place to make their habitat which is harmful to surroundings giving rise to various diseases.
- As irrigated water helps to lessen the inversion, sometimes it turns the area into a wet and sweaty surface. Immoderate irrigation clogs specific area by raising the water.
- While constructing the colossal irrigation waterways, people lose their residential and commercial lands.
- Not every time irrigation is affordable. The embryonic value of irrigation is quite inflated, and people end up paying a lot of imposed taxes.
- Irrigation becomes an obstacle in the path of free devours during the rainy season.
- Careless and over-irrigation can lead to salt efflorescence.

Irrigation Management

Irrigation water management is the act of timing and regulating irrigation water applications in a way that will satisfy the water requirement of the crop without the waste of water, soil, plant nutrients, or energy. It means applying water according to crop needs in amounts that can be held in the soil available to crops and at rates consistent with the intake characteristics of the soil and the erosion hazard of the site.

Management is a prime factor in the success of an irrigation system. Large quantities of water, and often large labor inputs, are required for irrigation. The irrigator can realize profits from investments in irrigation equipment only if water is used efficiently. The net results of proper irrigation water management typically:

- Prevent excessive use of water for irrigation purposes.
- Prevent irrigation induced erosion.
- Reduce labor.
- Minimize pumping costs.
- Maintain or improve quality of ground water and downstream surface water.
- Increase crop biomass yield and product quality.

Tools, aids, practices, and programs to assist the irrigator in applying proper irrigation water management include:

- Applying the use of water budgets or balances to identify potential water application improvements.
- Applying the knowledge of soil characteristics for water release, allowable irrigation application rates, available water capacity, and water table depths.
- Applying the knowledge of crop characteristics for water use rates, growth characteristics, yield and quality, rooting depths, and allowable plant moisture stress levels.
- Water delivery schedule effects.
- Water flow measurement for on field water management.
- Irrigation scheduling techniques.
- Irrigation system evaluation techniques.

Irrigation Water Management Concepts

The simplest and basic irrigation water management tool is the equation:

$$QT = DA$$

Where:

Q = flow rate (ft3 /s)

T = time (hr.)

D = depth (in)

A = area (acres)

For example, a flow rate of 1 cfs for 1 hour = 1 inch depth over 1 acre. This simple equation modified by overall irrigation efficiency, can be used to calculate the daily water supply needs by plants, number of acres irrigable from a source, or the time required to apply a given depth of water from irrigation well or diversion. Typically over 80 percent of IWM concerns can be at least partly clarified by the application of this equation.

When to Irrigate

This is dependent on the crop water use rate, (sometimes referred to as irrigation frequency). This can be determined by calculation of ETc rate for a specific crop stage of growth, monitoring plant moisture stress levels, monitoring soil water depletion and rainfall events. Applied irrigation water should always be considered supplemental to rainfall events. The irrigation decision maker should leave between 0.5 and 1.0 inch of available water capacity in the soil profile unfilled for storage of potential rainfall. Rainfall probability during a specific crop growing period and the level of risk to be taken must be carefully considered by the irrigation decision maker.

Water Measurement

A key factor in proper irrigation water management knows how much water is available to apply or is being applied to a field through an irrigation application system. Many devices are available to measure pipeline or open channel flows. Too many irrigators consider water measurement a regulation issue and an inconvenience. Typically less water is used where adequate flow measurement is part of the water delivery system.

Irrigation Scheduling

The determination of when and how much to apply requires a knowledge of the available water capacity (AWC) of the soil, the management allowed depletion (MAD) or plant stress level for the specified crop, the crop peak consumptive use, crop rooting depth, and the critical periods in the growing season when the crop should not be stressed.

Most crops should be irrigated before more than half of the available moisture in the crop root zone has been used. Some crops, however, are thought to do better at higher moisture levels (less moisture deficiency at time of irrigation). Generally, however, the need for irrigation is doubtful until the moisture deficit approaches one-third of the AWC of the crop root zone.

Irrigation must begin in time so that the entire irrigated area can be covered before the available moisture level in the last portion of the field to be irrigated reaches a point to cause unfavorable moisture stress of the crop. This aspect of management is crucial for systems that may need several days to irrigate the entire field area. Examples of such systems may be traveling gun systems, hand move lateral systems, and traveling lateral systems. One of the most effective ways of determining when and how much water to apply is to measure or estimate the soil water content as discussed previously. Measurements should be made in that part of the soil from which plant roots extract their moisture and according to the moisture extraction pattern of the particular crop. There are other methods being developed to determine when to irrigate, but measuring soil moisture is the most effective method in use now.

Measurements should be taken weekly in spring and fall and more frequently during the hot weather and critical growth periods of the crop. The irrigator may be able to reduce the frequency of readings after he or she has become familiar with the pattern of moisture depletion. To accurately predict moisture levels, measurements should be taken and recorded regularly, regardless of the time of year or the stage of crop growth. Comparison of yearly records with crop yields helps the irrigator to improve his or her management of the irrigation system.

Irrigation Scheduling Methods

In New Jersey the following methods are used to schedule when and how much water to apply. These include: soil and crop monitoring methods; the checkbook method; and computer assisted methods. Growers are recommended to use either of these methods depending on their management preference.

Soil moisture content should be monitored to determine if irrigation is needed based on predetermined critical levels for certain crops. The crop stress index method measures plant condition and compares that status to a well-known reference for a well-watered plant condition. Soil moisture monitoring before, during, and after the crop growing season is perhaps the most accurate irrigation scheduling tool.

Monitoring actual soil moisture is like receiving your bank statement from the bank, it affirms or cautions you when an error may exist or other adjustments may be needed. It is used together with the Checkbook Method of irrigation scheduling. This method has proven very useful for scheduling irrigations by providing a running account of available moisture in the effective root zone. Similar to bank account records of deposits (irrigation and rainfall) and withdrawals (evapotranspiration), the account balance provides the irrigator with information as to when to irrigate and how much water to apply. The method requires a daily recording of rainfall, estimated consumptive use, net irrigation amounts, and moisture balance throughout the growing season of the irrigated crop.

With internet access, growers can enroll in the South Jersey RC&D Weather Station Network which offers computerized irrigation scheduling using real time climate data to compute daily crop evapotranspiration. Using this service, the computer facilitates irrigation water management data as well as record keeping on the farm. (Computer assisted checkbook method).

Irrigation scheduling utilizes two important principles:

- When an adequate supply of available moisture is present in the effective root zone, the rate of consumptive use by a given crop depends primarily on. The stage of growth and climatological conditions.

- When the moisture content of the effective root zone is known at any given time, the moisture content at any later time can be computed by crediting moisture gained from effective rainfall or irrigation and subtracting the daily moisture withdrawals during the elapsed time.

To apply the above principles, the following requirements are essential:

- Soil with good internal and surface drainage.
- An adequate irrigation system and water supply.
- Daily consumptive use values for the crop.
- Accurate total available moisture values.
- Determination of the effective root zone of the crop.
- Measurements of effective rainfall and irrigation applications at the site.
- Available soil moisture maintained above the lower limit of withdrawal (25 to 30 percent of the total available moisture.) It is desirable to make periodic soil moisture checks to determine actual available moisture.

Equipment Required:

- Wedge-shaped plastic rain gage, 2- by 2-1/2 inch minimum top opening.
- Record book.
- Means of measuring average application rate of irrigation emission devices.

Rain Gauge Installation

To obtain a measurement of the rainfall in the irrigated area, the rain gauge should be set on a post located in field or in an open area adjacent to the field being irrigated. Its distance from nearby obstructions that might affect the catch, such as trees or buildings, should.be at least twice the height of the controlling obstruction. The gauge should be 3 to 4 feet above ground level with its top at least 3 inches above the top of the post.

The amount of rainfall in the gauge should be read at the lowest point of the water surface (meniscus). The reading should be made as accurately as possible, preferably to the nearest one-hundredth of an inch. After the rainfall amount has been recorded, the gauge should be emptied and reset in its holder.

Use of the Moisture Balance Sheet for Scheduling Irrigation

Record all available pertinent information in the heading:

- Determine the total available moisture holding capacity of the effective root zone and record it in the heading of the form.

- Determine the balance when irrigation is to begin, (MAD). The value frequently used is 50 percent of AWC.

Recording Procedure

- On a date near the normal planting date, or start of growth for forage crops and tree fruits, when the field is at field capacity, enter the total available moisture-holding capacity of the soil in the soil water content column on the Scheduling Worksheet.
  - When the soil moisture content is below field capacity at the time the moisture balance sheet is started, the available moisture remaining at that time will have to be estimated or measured. If the moisture percentage at field capacity has been previously determined for the field, measurement of present moisture percentage is all that is needed to provide an accurate starting value.
  - If field measurement of moisture percentage at field capacity has not been possible, begin the account with an estimated value. At the first opportunity after adequate rain or irrigation, obtain the field capacity moisture percentage and correct the account accordingly.
  - If at the time of plant emergence, a week or more has passed since the soil was at field capacity, and land preparation and seeding operations have ensued, the moisture balance may be estimated as being about 0. 5 inch below the field capacity value.
  - Early season moisture losses prior to plant emergence may be estimated by using an evaporation rate from the soil surface of 0.03 inch per day for the first week after the soil has been wetted to field capacity. An evaporation rate of 0.03 inch per week may be used for the remaining period. For bare or tilled soil conditions during May or later season plantings, when the mean daily temperature is above 63 degrees F, evaporation from the top 6 inches of soil may be estimated to average 0.1 inch per day for the first 5 days, and 0.05 inch per week thereafter.
  - Moisture losses may be estimated by measuring the depth of the dried surface layer and multiplying this depth by the available water holding capacity of the soil per inch of depth.
  - When a sod or cover crop is plowed under shortly before the planting date, the moisture balance may be considerably less than field capacity. Unless total rainfall between plowing and planting has brought the moisture level to field capacity or above, the moisture content of the soil should be measured at the start of record keeping.
- At the end of the first day of moisture accounting:
  - Estimate crop consumptive use, ETc: Real time weather data from SJRC&D Weather Station Network which calculates ETc; Irrigation Water Requirements Software Program (uses historical weather data and crop coefficients) or evapotranspiration tables for North, South, and Central NJ.

   - Record the amount of rainfall or the net amount of irrigation applied. Amounts should be recorded to nearest 0.01 inch.
   - Starting with the preceding day's balance, subtract the Et use value. If there has been rainfall exceeding 0.2 inches, or irrigation, add these values to the daily moisture balance to obtain the new daily balance. If the total exceeds the value for Available Moisture at field capacity in the root zone, enter the Available Moisture recorded in the heading of the form. If the total exceeds the Available Moisture value at field capacity by 0.5 inch or more, continue the full moisture balance value for two succeeding days. During this 2-day period the plants will be using the free water in the root zone.
- Continue the same process as outlined above at the end of each day during the growing season until the crop has reached the desired maturity level.

When the daily moisture balance is allowed to drop below 50 percent of the total available moisture value, soil moisture tensions will increase to the point where it becomes increasingly difficult for the plants to obtain the moisture they need for continuous vigorous growth. In such cases, daily evapotranspiration will be reduced.

If for some reason irrigation is not accomplished when the daily moisture balance has reached the 50 percent level, Et values must be reduced. The actual reduction in Et rates varies with the kind of crop and is difficult to define; however, the following factors can be used as a guide for estimating probable daily Et rates:

Moisture level, percent	Factor
40-50	0.8
30-40	0.6
20-30	0.2

The net application is determined by:

- Obtaining the difference between the recorded daily balance prior to the start of irrigation and the available moisture at field capacity in the root zone.
- Adding to this value the estimated Et for that day.
- Making proper allowance for any effective rainfall that may have occurred since recording the last daily balance. The gross application is determined by dividing the net application amount by the expected efficiency of the irrigation method used on the particular field.

Variations in the procedure may be used to fit individual situations. The time required to apply the required gross application is determined by dividing the gross application amount by the application rate of the irrigation system as determined from an on-site evaluation of the irrigation system.

Location of Soil-Moisture Devices

The selection of soil moisture measurement stations is important. The stations should be located so that average soil moisture conditions in the root zone of the crop are measured. Excess water from leaks in pipe joints, low spots in a field, etc., should not be allowed to come in contact with the measurement station. High spots with excessive water runoff should not be chosen because the soil profile in this area will not represent average root zone conditions. Average soil and slope conditions in the field should be represented in station locations. Measurements should be made at other locations as indicated by any critical condition in the soil, such as an area that dries out first. It is a good practice to have at least two measurement stations in each critical area and two or three stations in areas that are typical of the field. This information provides direction for adjusting the amount and frequency of irrigation for different parts of the field or for different periods in the growing season. Measurement Depth:

- In uniformly textured soils, one measurement should be made at the midpoint in each quarter of the root zone. For shallow rooted crops (maximum 3 foot deep root zones), it is probably desirable to take three measurements. As an example, in a 24-inch root zone, measurements may be taken from the 6-, 12-, and 18-inch depths.
- In stratified soils, one measurement should be taken from each textural strata. It may not be necessary to take a measurement in very thin layers when this thin layer can be lumped with another layer for estimating soil moisture. Where the strata is thick, a sample should be taken in 1-foot increments as a minimum. Thickness of the strata should be noted.
- The crop root depth for annual crops changes through the early part of the growing season. Measurements should be made in the soil profile according to the current depth of the majority of the crop roots.

Location in Relation to Plants: For row crops, locate the measurement in the crop row as near the plants as possible. For complete cover crops, such as alfalfa and small grains, locate in representative soil and slope areas of the field. For trees, locate about 4 to 6 feet from the trunk, but inside the tree drip line. Location in Relation to Irrigation Systems:

- Lateral move (side roll or hand move) sprinklers: locate measurement stations halfway between adjacent sprinkler heads and 10 to 15 feet from the lateral.
- Center pivot sprinklers: locate measurement stations at about two-thirds of the total lateral distance from the pivot.
- Solid set sprinklers: locate measurement stations where the diagonals from four adjacent sprinkler head cross.
- Trickle systems: locate in the wetted ball in the root zone.
- Furrows and corrugations: measurement stations should be located in about the center of the furrow or corrugation bed near the plants.

Location in Field for Sprinkler or Trickle Irrigation System: Sprinkler and trickle irrigation systems lose pressure down the lateral due to friction loss throughout the lateral, so sprinkler heads farthest from the main lines put out the least irrigation. Locate measurement stations as follows:

- 50 to 100 feet downstream from the beginning of the lateral.
- 50 to 100 feet upstream from the distal end of the lateral.
- At least one measurement close to the center of the lateral. For laterals 1,320 feet to 2,640 feet long, locate two measurement stations at the one. For shorter laterals, one center third points along the lateral measurement should be adequate.

Irrigation Engineering

Irrigation engineering is a kind of civil engineering which includes the study of controlling and harnessing the various natural sources of water. It studies the irrigation and agriculture in-depth. Flooding is the basis of agriculture and the only way by which water is supplied to the crops. Irrigation engineering is a field of engineering which would discuss the uses of irrigation in crop production and how new principles and technologies can be used and applied for irrigational practices. It is essential to study irrigation engineering as it helps in determining future irrigation expectations and impacts of irrigation in the agriculture field.

There are different types of irrigation techniques, and Irrigation engineering includes the study of the design of different irrigation systems and analyzes the efficiency of each of the system and list their advantages and disadvantages.

Irrigation Engineering is very important for designing and planning of cost-effective and efficient irrigation systems. There are various benefits as well as risks of the irrigation system. For example, it increases the groundwater level and aid to the cultivation and also protects a country from a famine situation. However, there are some ill-effects too of irrigation such as it creates a breeding place for the mosquitoes. Also, valuable land may be submerged because of the construction of dams, and barrages. The benefits are quite more than the ill-effects of the irrigation system.

References

- Irrigation: byjus.com, Retrieved 12, February 2020
- Types-methods-importance-irrigation: conserve-energy-future.com, Retrieved 03, April 2020
- Advantages-and-disadvantages-of-irrigation, water-resource-engineering-268: civiltoday.com, Retrieved 26, August 2020
- What-is-irrigation-engineering: airtract.com, Retrieved 10, March 2020

Chapter 2

Soil and Water Relations

Soil is a complex mixture of minerals, organic matter, air, water and organisms that support life. Soil moisture is the content of water in the soil. Water present in the soil is classified into hygroscopic water, capillary water and gravitational water. The topics elaborated in this chapter will help in gaining a better perspective about the soil-water relations.

The plant needs soil and water for its growth. The soil anchors the plant and allows the root system to grow. Soil consists of a three-phase system – solid, liquid and gas. The complex mass of organic and mineral matter acts as a matrix consisting of a number of pores of various sizes. The void space within the particles is known as soil pore space. The pores of the soil hold moisture attached to the soil particle by surface tension. If water is present in amounts more than that can be stored, the soil expels air. On the other hand, in dry soils, the water molecules are present as a thin film around the soil particles.

Classification of Soil Water

Water occurs in varying proportions in the soil pores. Based on the physical factors, the water held in the pores can be defined as:

- When a soil sample is saturated with water and allowed to drain the amounts that are present in excess, the gravitational water is drained off easily and rapidly from the root zone. Gravitational water is not available for use by plants.
- The amount of water retained in the soil by surface tension after the draining of gravitational water is known as capillary water. Capillary water is the main source of water for plant growth.
- Hygroscopic water is the moisture content absorbed and retained by dry soil as a thin film over the soil particles. Neither this water is available to plants.

Importance of Water in Plant Growth

Water plays a key role in plant growth. The most important functions of water are listed below:

- Water constitutes 85-90% of the body weight of younger plants and 20-50% of the weight of mature plants.
- Oxygen and hydrogen required for carbohydrate synthesis during the photosynthesis is provided by water.
- Nutrients required for the plant growth are transported from soil to the plant with the help of water.

- Food synthesized in the leaves gets dissolved in the water and then distributed throughout the plant body.
- Water is essential for optimal transpiration. Transpiration helps maintain the plant temperature through dissipation of heat.

Soil-Plant-Atmosphere Continuum

The pathway through which water moves from the soil to the atmosphere via plant is known as the soil-plant-atmosphere continuum. The atmosphere has low water potential while a relatively higher potential occurs inside the leaves. This creates a diffusion gradient across the pores in stomata of the leaves leading to vaporization of water molecules present in the leaves. As water vapor transpires out of the leaves, it is replaced with water molecules maintained inside the leaves at saturation vapor pressure. Xylem cells of plant pull water from the roots towards the leaf.

Soil

Soil is the loose surface material that covers most land. It consists of inorganic particles and organic matter. Soil provides the structural support to plants used in agriculture and is also their source of water and nutrients. Soils vary greatly in their chemical and physical properties. Processes such as leaching, weathering and microbial activity combine to make a whole range of different soil types. Each type has particular strengths and weaknesses for agricultural production.

Physical Characteristics of Soil

The physical characteristics of soils include all the aspects that you can see and touch such as:

- Texture
- Color
- Depth
- Structure
- Porosity (the space between the particles)
- Stone content

Good soil structure contributes to soil and plant health allowing water and air movement into and through the soil profile. Soil stores water for plant growth and supports machine and animal traffic. While some soils are naturally better structured than others, some physical characteristics of soils can be changed with good management.

It is important to monitor the physical characteristics of soil to understand soil condition. It is also important to ensure that management practices are not contributing to the decline of the soil. An example of this is excessive traffic causing compaction and reducing the amount of macropores, or spaces between the aggregates, therefore reducing the amount of air and water into and through the soil.

Soil Color

Soil color can indicate the organic matter content of soil; the parent material soil is formed from, the degree of weathering the soil has undergone and the drainage characteristics of the soil. The color of the soil is the main indicator of how soils drain.

Table: Soil color and indications.

Soil color	Indication
Dark brown	High organic matter content
Black	Humus
Red	Presence of iron Phosphorous may be less available to the plant Free draining
Yellow	Moist conditions Restrictive drainage Less weathering
Grey, Blue/green hues	Poor drainage Waterlogging

Lighter colored soils can generally indicate low fertility for example white sands. While darker soils (like black clays) are quite fertile. There is a large range in between.

Determine Soil Drainage

The drainage of a soil is an important characteristic to assess, as many plants prefer well-drained soils. If a soil is poorly drained, sufficient oxygen cannot get to the plant roots which can stunt or kill the plant. Soils that are very well drained can limit plant capture of water in drier environments or in dry years due to insufficient water holding capacity.

Other important indicators are:

- Texture of the soil.
- Presence of buckshot and stones.
- Dispersibility and friability of the soil.

Inorganic Component of Soils

Inorganic material is the major component of most soils. It consists largely of mineral particles with specific physical and chemical properties which vary depending on the parent material and conditions under which the soil was formed.

It is the inorganic fraction of soils which determines soil physical properties such as texture. This has a large effect on structure, density and water retention.

Soil Texture

The texture of soil is a property which is determined largely by the relative proportions of inorganic particles of different sizes.

In Australia, the following sizes are used to describe the inorganic fraction of soils:

- Gravel: Particles greater than 2mm in diameter.
- Coarse sand: Particles less than 2mm and greater than 0.2mm in diameter.
- Fine sand: Particles between 0.2mm and 0.02mm in diameter.
- Silt: Particles between 0.02mm and 0.002mm in diameter.
- Clay: Particles less than 0.002mm in diameter.

Sand

Quartz is the predominant mineral in the sand fraction of most soils. Sand particles have:

- A relatively small surface area per unit weight.
- Low water retention.
- Little chemical activity compared with silt and clay.

Silt

Silt has a relatively limited surface area with little chemical activity. Soils high in silt may compact under heavy traffic. This affects the movement of air and water in the soil.

Clay

Clays have very large surface areas compared with the other inorganic fractions. As a result, clays are chemically very active and able to hold nutrients on their surfaces. These nutrients can be released into soil water to use by plants. Like nutrients, water also attaches to the surfaces of clays but this water can be hard for plants to use.

There are many different types of clays. Clays are distinguished from sand and silt by their ability to swell and retain a shape they have been formed into — as well as by their sticky nature.

Soil Textural Class

The relative proportion of sand, silt and clay particles determines the physical properties of soil including the texture. The surface area of a given amount of soil increases significantly as the particle size decreases. Consequently, the soil textural class also gives an indication of soil chemical properties.

The exact proportions of sand, silt and clay in a soil can only be determined in a laboratory. However, a naming system has been developed to approximately describe the relative proportions. This classification of soil can be undertaken in the field where particular properties indicate possible textural classes.

To estimate texture in the field, crush a small sample of soil (10 to 20g) in one hand. After removing any gravel or root matter, work the soil in the fingers to break down any aggregates. With the sample moist but not sticky, the textural class can be estimated by the feel of the sample between the fingers.

Textural Class Descriptions for Soil

A simple way to determine a soil texture and its characteristics is by hand texturing. When texturing soil it is important to understand the behaviour feel, color, sound and cohesiveness of the soil, which is achieved by making a bolus (wetting the soil and forming a ball). For example, a sandy loam will only just stick together (slightly coherent) and there will be noticeable sand grains which can be seen and felt and heard if you place the bolus close to your ear and squeeze it.

It is then important to form a ribbon from the bolus to determine the clay content of the soil. The longer the ribbon, the higher the clay content. The length of the ribbon is measured against a ruler and along with the behaviour of the soil can be compared with the descriptions on the soil texture table. This table will help you to assess soil texture.

Table: Guide to common soil textures.

Texture grade	Behaviour of moist bolus (ball formed in palm of hand)
Sand	Coherence, nil. Single sand grains adhere to fingers. If you press the bolus between your fingers, holding close to your ear, you will hear the sand grains rubbing against each other.
Loamy sand	Slight coherence. Discolors fingers with dark organic stain. Ribbon length 1.0cm.
Clayey sand	Slight coherence; sticky when wet. Many sand grains stick to fingers. Discolors fingers with clay stain. Ribbon length 1.0cm.
Sandy loam	Bolus just coherent but very sandy to touch. Ribbon length 1.3 to 2.5cm. Can hear sand grains (see Sand description).
Fine sandy loam	Bolus coherent. Sand can be felt and heard when manipulated. Ribbon length 1.3 to 2.5cm.
Light sandy clay loam	Bolus strongly coherent but sandy to touch. Ribbon length 2 to 2.5cm.
Loam	Bolus coherent and spongy. Smooth feel, may be greasy. Ribbon length 2.5cm.
Loam fine sandy	Bolus coherent and slightly spongy. Fine sand can be felt and heard when manipulated. Ribbon length 2.5cm.
Silt loam	Coherent bolus, very smooth to silky when manipulated. Ribbon length 2.5cm.
Sandy clay loam	Strongly coherent bolus sandy to touch. Medium sand grains visible. Ribbon length 2.5 to 3.8cm.
Clay loam	Coherent plastic bolus, smooth to manipulate. Ribbon length 4 to 5cm.
Silty clay loam	Coherent smooth bolus, plastic and silky to touch. Ribbon length 4 to 5cm.
Fine sandy clay loam	Coherent bolus, fine sand can be felt and heard. Ribbon length 4 to 5cm.
Sandy clay	Plastic bolus, fine medium sands can be seen, felt or heard in clay matrix. Ribbon length 5 to 7.5cm.

Silty clay	Plastic bolus, smooth and silky to manipulate. Ribbon length 5 to 7.5cm.
Light clay	Plastic bolus, smooth to touch; slight resistance to shearing between thumb and forefinger. Ribbon length 5 to 7.5cm.
Light medium clay	Plastic bolus, smooth to touch, slightly greater resistance to ribboning. Ribbon length 7.5cm.
Medium clay	Smooth plastic bolus, handles like plasticine. Has some resistance to ribboning. Ribbon length 7.5cm.
Heavy clay	Smooth plastic bolus, handles like stiff plasticine. Has firm resistance to ribboning. Ribbon length 7.5cm or more.

It should always be remembered that soil texture often varies with depth and that the properties of the topsoil are affected by the properties of the subsoil.

Structure

Structure is the arrangement of soil particles and pore spaces between. Soil with a structure beneficial to plant growth, has stable aggregates between 0.5 and 2mm in diameter. Such soils have good aeration and drainage.

Chemical Properties

The inorganic minerals of soils consist primarily of silicon, iron and aluminum which do not contribute greatly to the nutritional needs of plants. Those in the clay fraction have the capacity to retain nutrients in forms which are potentially available for plants to use.

Organic Component of Soil

The organic matter of soil usually makes up less than 10% of the soil. It can be subdivided into living and the non-living fractions. The non-living fraction contributes to the soil's ability to retain water and some nutrients and to the formation of stable aggregates.

Organic Matter Fraction of Soils

The organic matter fraction of soils comes from the decomposition of animal or plant products such as faeces and leaves. Soil organic matter contributes to stable soil aggregates by binding soil particles together.

Plants living in soil continually add organic matter in the form of roots and debris. Decomposition of this organic matter by microbial activity releases nutrients for the growth of other plants.

The organic matter content of a soil depends on the rates of organic matter addition and decomposition. Soil microorganisms are responsible for the decomposition of organic matter such as plant residues. Initially, the sugars, starch and certain proteins are readily attacked by a number of different microorganisms. The more resistant structural components of the cell wall decompose relatively slowly. The less easily decomposed compounds, such as lignin and tannin, impart a dark color to soils containing significant organic matter content.

The decomposition rate of organic materials depends on how favorable the soil environment is for microbial activity. Higher decomposition rates occur where there are:

- Warm, moist conditions
- Good aeration
- A favorable ratio of nutrients
- A pH near neutral
- Freedom from toxic compounds

Soil Organisms

The soil contains numerous organisms ranging from microscopic bacteria to large soil animals such as earthworms. The soil microorganisms include:

- Bacteria.
- Fungi.
- Actinomycetes.
- Algae.
- Protozoa.
- Nematodes.

The diversity of soil organisms can both assist and hinder plant growth. Beneficial activities include:

- Organic matter decomposition.
- Nitrogen fixation.
- Transformation of essential elements from one form to another.
- Improvement in soil structure through soil aggregation.
- Improved drainage and aeration.

Under some circumstances soil organisms compete with plants for nutrients. Bacteria are the smallest and most numerous microorganisms in the soil. They make an important contribution to organic matter decomposition, nitrogen fixation and the transformation of nitrogen and sulphur.

The fungi and actinomycetes contribute beneficially to organic matter decomposition. The group of large soil animals includes earthworms, which incorporate organic matter into the soil as well as improving aeration and drainage by means of their channels. Some soil fungi, nematodes, and insects feed on roots and lateral shoots to the detriment of plants.

Soil Composition

Soil composition is an important aspect of nutrient management. While soil minerals and organic matter hold and store nutrients, soil water is what readily provides nutrients for plant uptake. Soil air, too, plays an integral role since many of the microorganisms that live in the soil need air to undergo the biological processes that release additional nutrients into the soil.

The basic components of soil are minerals, organic matter, water and air. The typical soil consists of approximately 45% mineral, 5% organic matter, 20-30% water, and 20-30% air. These percentages are only generalizations at best. In reality, the soil is very complex and dynamic. The composition of the soil can fluctuate on a daily basis, depending on numerous factors such as water supply, cultivation practices, and/or soil type.

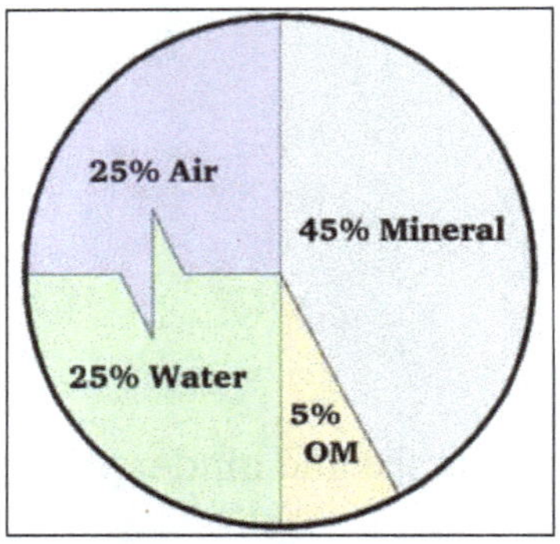

Approximate composition of soil.

The solid phase of soil, which includes minerals and organic matter, are generally stable in nature. Yet, if organic matter is not properly managed, it may be depleted from the soil. The liquid and gas phases of the soil, which are water and air respectively, are the most dynamic properties of the soil. The relative amounts of water and air in the soil are constantly changing as the soil wets or dries.

Soil Profile

The soil profile is one of the most important concepts in soil science. It is a key to understanding the processes that have taken in soil development and is the means of determining the types of soil that occur and is the basis for their classification. The soil profile is defined as a vertical section of the soil from the ground surface downwards to where the soil meets the underlying rock. The soil profile can be as little as 10 cm thick in immature soils and as deep as several meters in tropical areas where the climate is conducive to rapid alteration of the underlying rock to form soil. In temperate areas, the soil profile is often around a metre deep and in arid areas somewhat shallower than this.

Virtually all soil profiles are composed of a number of distinctive layers, termed horizons, interpretation of which is the key to understanding how the soil has formed. Most soils will have three or more horizons. Soils that have not been cultivated will normally have L, F and H layers at the surface. These layers largely represent different degrees of decomposition of organic matter, the L layer representing the litter layer formed of recognizable plant and soil animal remains, the F layer below, the fermentation layer, usually consisting of a mixture of organic matter in different stages of decomposition, and the H layer, the humus layer, consisting largely of humified material with little or no plant structure visible. Below these, and in cultivated soils occupying the surface layer, is the A horizon composed of a more or less intimate mixture of mineral and organic matter. The A horizon is often referred to as the 'ploughed layer' in cultivated soils. It is an important part of the soil because it is a source of plant nutrients and contains the majority of plant roots. The A horizon may lie directly on the B horizon or, in well-developed soils, there may be an intermediate leached horizon, termed E or A2, depending on the nomenclature system used. The E/A2 horizon is usually paler in color than the horizons above and below because it is a horizon that has been subject to leaching and loss of components compared to the A and B horizons.

The B horizon is the horizon most widely used to identify soil types. Its morphology is important in supporting the classification of soils. In some soils the B horizon results purely from the weathering of the underlying rock whereas in other soils this weathering is supplemented by the translocation of materials from overlying horizons. Thus the B horizon needs to be inspected carefully in order to understand the genesis of the soils. B horizons may have a number of different subscripts indicative of the nature of the materials that have moved into the horizon, e.g. Bh indicates the translocation of humus into the horizon, Bs, the translocation of sesquioxides. These subscripts will vary according to the nature of the soil component that has accumulated but also with the system of nomenclature of soil types.

Below the B horizon is the C horizon. This latter horizon is often consistent with the parent material and may have been little altered from the material in which the soil originally formed.

Most soils have A, B and C horizons. Some, generally weakly developed, soils may have A horizons lying directly on C horizons.

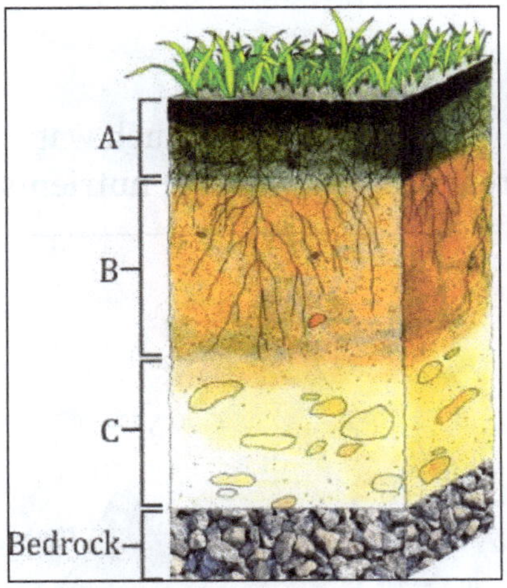

Soil Texture

Soil texture refers to the composition of the soil in terms of the amounts of small (clays), medium (silts), and large (sands) size particles. The primary particles of sand, silt, and clay make up the inorganic solid phase of the soil. These particles often become aggregated together with each other and other parts of the soil, most importantly soil organic matter.

In general, the size of soil particles and their spacing determine how much water can flow through the soil. The larger the spacing, or pore size, the greater the infiltration rate. Thus, sandy soils will have high infiltration rates because pore sizes are large and there are no finer materials to block the pores. The soil texture also influences how much heat and nutrients will be stored in the soil profile.

Sandy soils drain better than soils that are clay rich. In general, the smaller the soil particle size distribution, the slower it will drain. Sometimes silt may have the same particle size distribution as clay, but clay will retain more water for longer periods of time than silt. This can be explained by the shape of the soil particles. Clay particles are planar whereas silt particles are spherical. Water basically gets stuck between the planar plate shaped clay particles and thus slows the flow of water.

An easy way to help determine what type of soil you have is to simply feel it to determine the texture and thus what the primary makeup of the soil is. Grab a baseball size portion of the soil in your hands and wet the soil with water, working the moist soil with your hands. The stickier it is, the more clay there is. The "soapier" the soil feels the higher the silt content. Grittiness is indicative of sand.

In the United States, twelve major soil texture classifications are defined by the USDA. The twelve classifications are sand, loamy sand, sandy loam, loam, silt loam, silt, sandy clay loam, clay loam, silty clay loam, sandy clay, silty clay, and clay. Soil textures are classified by the fractions of each soil component (sand, silt, clay) present in the soil. Classifications are typically named for the primary constituent particle size or a combination of the most abundant particles sizes, e.g. "sandy clay" or "silty clay". A fourth term, loam, is used to describe equal properties of sand, silt, and clay in a soil sample, and lends to the naming of even more classifications, e.g. "clay loam" or "silt loam".

Texture is important because it influences:

- The amount of water the soil can hold.
- The rate of water movement through the soil.
- How workable and fertile the soil is.

For example, sand is well aerated but does not hold much water and is low in nutrients. Clay soils generally hold more water, and are better at supplying nutrients.

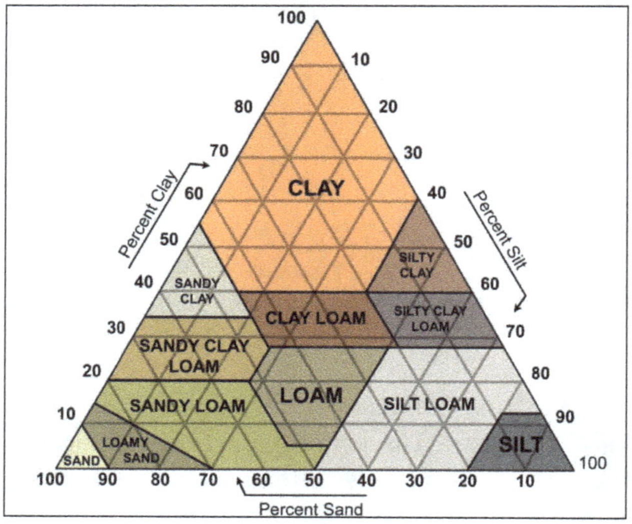

Soil Structure

Soil structure has a major influence on soil behaviour. A 'well structured' soil has plenty of living spaces, storage spaces, doorways, and passages (for utilization by water, gases, nutrients, roots and a vast array of organisms). A poorly structured soil is much less endowed, and much less productive.

Soil structure is the arrangement of pores and fissures (porosity) within a matrix of solid materials (soil particles and organic matter). The solid materials bond and aggregate to give the pores and fissures. The quantity, distribution and arrangement of pores determines water holding capacity, infiltration, permeability, root penetration, and, respiration.

Soil Structure Close-Up

Only about 50% of soil is solid material. The remainder is pore space. It is in these spaces that the action happens. Water is stored there. Organisms live there. Organic matter and nutrients accumulate there.

The diagram (magnified about 20 times) demonstrates how solids and pores might arrange in soil to give a porosity of 50 %.

Small pores within the aggregates provide storage and refuge. The larger pores (and fissures) between the aggregates are the pathways for liquids, gases, roots and organisms.

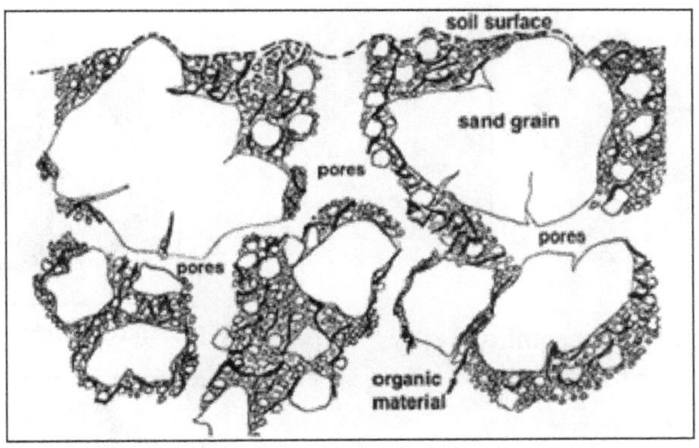

The diagram is also useful to imagine how various management actions may change soil structure. What would happen if a weighted wheel passed over this surface? What would happen if a Tyne were dragged through it? What would happen if half of the organic was removed?

Importance

Soil is like a city. The structure and layout of both determine how things happen, the rate at which they happen, and the capability to keep them happening.

The following characteristics are used to help evaluate the ability of any soil to perform well (or otherwise):

- Porosity (to represent aeration, water storage capacity, plant wilting point and drainage).
- Permeability (to represent infiltration, drainage and respiration).

- Bonding and aggregation (to represent how the solids group together and the construction materials used).
- Soil strength (to represent toughness and resilience of structures).
- Friability, tillage and trafficability (to represent how soils behave with mechanical disturbance).

Types

Soil material fits and binds together in many different ways. With some, the bonding is very weak, in others very strong. With some, the size of aggregates is very fine, in others coarse and large. With some the aggregates are dense containing few pores, in others quite open with plenty of pores. There are six broad categories of soil structure.

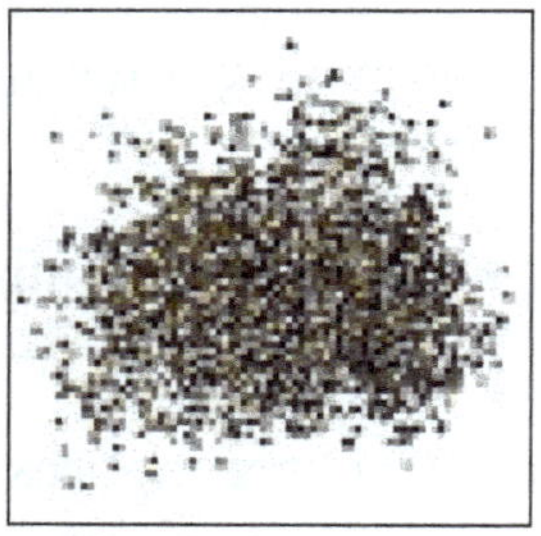

Granular (high permeability)

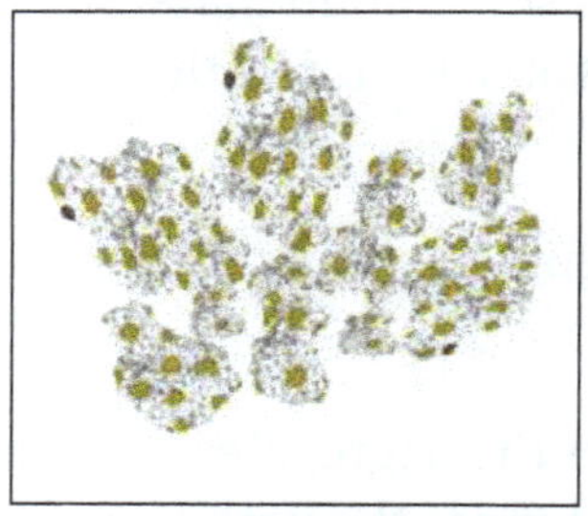

Aggregated (high permeability)

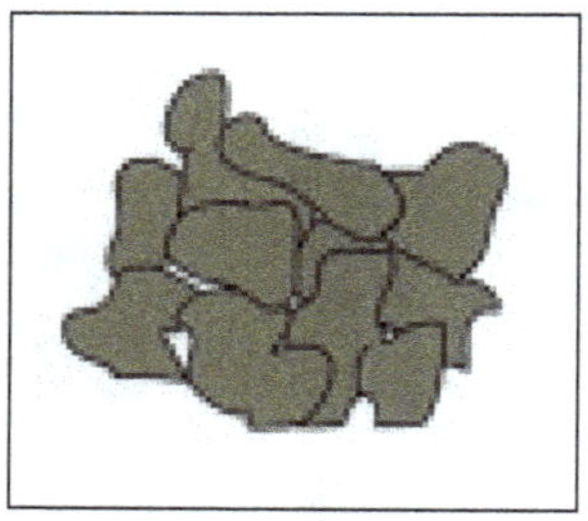

Blocky (moderate permeability)

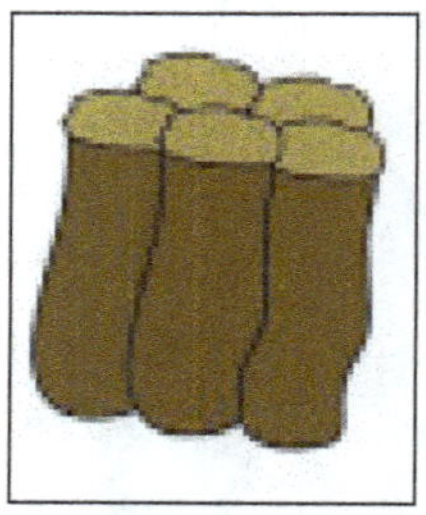

Columnar/prismatic (moderate permeability)

Platey (low permeability)

Massive (low permeability)

Indicators of Damaged Soil Structure

- Root restriction
- Compacted layers
- Plough pans
- Surface crusting

Benefits of Understanding Soil Structure

- Tilth and ease of tillage (time and costs).
- Trafficability (go and no-go periods).
- Irrigation management (frequency and intensity).
- Runoff and water quality.
- Interpreting surface features.
  - Crusts, sealing, erosion and deposition.
- Root exploration – occupation of soil, root deformation, root health.
- Density (porosity and compaction).
- Aggregation (clodiness and stability).
  - Slaking and dispersion, water-stable aggregates.
- Aggregate strength (dry) and friability.
- Water retention properties and water movement.
  - Waterlogging.
- Providing stable foundations.

Protecting and Improving Soil Structure

- Know the territory.
- Characterize the soil.
- Know the "fixed" limitations of your soil such as soil type and texture.
- Work on the "manageable" limitations, e.g.:
  - Aggregate stability.
  - Profile permeability.
- Prevent:
  - Compaction.
  - Plough pans.
  - Crusts.
  - Erosion.
- Manage the strengths, e.g. surface maintenance and protection organic matter cycling.

Soil Water

Soil water has traditionally been of interest because of its influence on plant growth and crop production as well as runoff processes. A growing plant must be able to balance the atmospheric demand for water with the amount it can extract from the soil. The soil water supply is alternately depleted through evapotranspiration and replenished by irrigation or precipitation. Today soil water is of increasing of concern as a medium through which chemicals may move and potentially harm surface or groundwater.

Soil Water Content

Soil holds water to the extent that they have porosity, and the water usually shares that pore space with air. Even "saturated" soil will usually have some air trapped within. The porosity of soil itself is quite variable, in response to both natural and agricultural practices. Soil water content, by weight, is calculated as:

$$\theta_w = \frac{W_{moist} - W_{dry}}{W_{dry}} = \frac{W_{moist}}{W_{dry}} - 1$$

Where:

θ_w= water content expressed on the basis of the dry weight of soil

W_{moist} = moist soil weight

W_{dry} = oven-dry soil weight

It is common to express soil water contents on a volumetric basis, i.e., the ratio of the soil water volume to the total soil volume. This is done by multiplying the water content on a dry weight basis by the ratio of the soil bulk density, ρ_b, and water density, ρ_w, as follows:

$$\theta_w = \theta_w \frac{\rho_b}{\rho_w}$$

Where θ = water content on a volume basis. The soil bulk density (or apparent density), ρ_b, is defined as the oven-dry weight of soil per unit volume, as it occurs in the field. In the following discussion, θ will be used to mean volumetric water content.

Soil Water Potential

Soil water content alone is not a satisfactory criterion for describing the availability of water to plants and attempts have been made to describe water availability in terms of the energy state of water. Initially, empirical measurements and relationships were developed, but these gave way to consideration of fundamental mechanisms and expressions. The soil-water-plant system is now treated as a continuous dynamic system where water moves through the soil to plant root surfaces, into roots, through the plant, and into the atmosphere along a path of continuously decreasing potential energy. The removal of soil water depends not only upon its amount and energy state, but also upon the ability of the plant to absorb water and the atmospheric demand for water from the plant.

About the beginning of the 20th century, soil water was arbitrarily classified into different forms such as gravitational water, capillary water, hygroscopic water, etc. These early groupings have been replaced by a fundamental concept referred to as soil water potential. Soil water does not occur in separable "forms" within the range of our interest, but does vary in the energy with which it is retained in the soil. The work per unit weight to move an infinitesimal amount of water from some reference state to a given point in the soil is known as the total soil water potential, h_T. The usual reference state, arbitrarily defined as having zero potential, is an open air-water interface at some specified elevation and air pressure. Energy must be expended to remove water from an unsaturated soil, so the soil water potential is less than the reference state and thus has a negative sign. The potential gradient, or rate of decrease of potential energy with distance, is the driving force causing soil water flow. Thus, soil water will move from a wet area where the potential is near zero, toward a dry region where the potential is lower (a larger negative value). The soil water pressure has dimensions of $[M/LT^2]$, and the equivalent potential h_T has dimension of length. Other definitions will follow resulting in dimensions of pressure and energy per unit mass.

The total soil water potential may be expressed as the sum of three component potentials:

$$h_T = h_g + h_p + h_a$$

Where:

h_T = total soil water potential.

h_g = gravitational potential.

h_p = matric or pressure potential.

h_a = pneumatic potential.

Gravitational potential, h_g, is the elevation. If z is the height above a defined reference plane, $h_g = z$. The value of hg can be positive or negative.

The value of h_p, the pressure (and matric) potential, can be positive or negative and is equal to the soil water pressure head (i.e., the pressure divided by the specific weight). If the soil water pressure is greater than the adjoining gas phase pressure, then hp will be positive. If the pressure of the soil water is less than the adjoining gas phase, h_p will be negative. This is due to the attraction of soil surfaces for water, the influence of soil pores, and the curvature of the soil water interface. For this situation, h_p is also called the matric potential. It is convenient to consider pressure potential as a continuous function of water content, which is positive in a saturated soil below the water table and negative in unsaturated soil. Since soil water potential is generally negative, it is often given a positive value and referred to as suction or tension.

The pneumatic potential, h_a, (energy per unit mass) may be expressed as $h_a = p_{sa} /(\rho\ g)$, where p_{sa} is the soil air pressure, ρ is the density of water, and g is gravitational acceleration. Usually the air pressure is considered to be uniform throughout the soil profile and the pneumatic potential is ignored in characterizing soil water flow.

Two other ways are used to define potential. These are energy per unit volume, $h_{T,v}$, and energy per unit mass, $h_{T,m}$. The dimensions of $h_{T',v}$ are pressure, and the relation to h_T, above, is:

$$h_{T,v} = \rho\, g h_T$$

Similarly, the relationship between h_T and $h_{T',m}$ is:

$$h_{T,m} = g h_T$$

Table: A useful conversion table taken from Hillel.

Soil Water Potential			Soil Water Suction		
Per Unit Weight (mm H2O)	Per Unit Mass (joules/kg)	Per Unit Volume (kPa)	Per Unit Weight (mm H2O)	Per Unit Volume (kPa)(bars)	
−102.0	-1	1	102.1	1.0	0.01
−1020	-10	-10	1020	10	0.1
−5100	-50	-50	5100	50	0.5
−10200	-100	-100	10200	100	1.0
−51000	-500	-500	51000	500	5.0

Design and Operation of Farm Irrigation Systems

To avoid confusion among the various expressions for soil water energy status, one must keep in mind that a low potential refers to dry soil and is a large negative number, while a high matric or pressure potential refers to a wet soil with a small negative value of h. A high potential would be −0.10 bar while a low potential would be −15 bars. On the other hand, low suction or tension refers to wet soil with a small positive suction value. High suction or tension means a dry soil and is a large positive number; i.e., a low suction is +0.10 bars and a high suction is +15 bars. The main incentive for introducing soil water potential, hT, is to describe flow relations based on spatial differences in hT. For a non-equilibrium system, flow will occur from a higher to a lower potential. Flow is influenced by additional factors and these factors are often included as additional components for hT. In particular, osmotic effects are often included as an additional osmotic potential component. The osmotic potential is a significant component in saline soils. For coupled flow processes consisting of flow due to osmotic gradients, temperature gradients, pressure gradients, and other gradients, it is not necessary to define a total potential which includes components for each independent part. In fact, Corey and Klute showed that the inclusion of chemical effects can lead to contradictions with the notion that flow occurs from regions of high to low potentials. (This does not complicate the formulation of coupled flows, but simply says that the same transport coefficients cannot be used for the independent gradient terms for each component.)

The Soil Water Retention Characteristic

As water is removed from a soil, the matric or pressure potential of the water remaining is decreased (algebraically, e.g., −1 is decreased to −10). If water is added to the soil, the matric potential is increased (such as, −10 to −1). A curve showing the functional relationship between matric potential and soil water content is known as the soil water characteristic or retention curve. Soil water is usually expressed as volumetric water content θ or as volumetric percentage of water. When the relationship is determined by drying a wet soil, the curve is known as either the desorption curve, water retention curve, or water release curve. When the relationship is determined as a dry soil wets, it is called the sorption or imbibition curve. The soil water characteristic is related in an indirect way to the pore size distribution.

Water is retained in the soil by a combination of the attraction of particle surfaces for water and the capillary action of water in the soil pores. The matric potential is related to the curvatures of the air-water interfaces, which in turn are affected by the soil pore geometry, the particle aggregation, and the soil water content. At high matric potentials (near zero), most of the soil pores are filled with water and the total porosity and pore size distribution greatly influence the water retained. Inasmuch as soil texture dominates the total porosity and pore size distribution, it has a marked effect on the soil water characteristic. In general, the higher the clay content of a soil, the higher will be the water content at any given potential. Soil aggregation, especially for fine textured soils, tends to increase the number of large pores. Thus, soil structure is important in the amount of water retained at high potentials. When the large pores empty, the water remaining in the soil is held in the smaller interaggregate pores and at the particle contact points. As the soil dries, the amount of particle surface area also affects the water retained, and this is strongly influenced by soil texture. Soil compaction also influences the water characteristic because compaction results in smaller pores, reduced total porosity, and increased interparticle contact in a given soil volume It is usually the larger pores that are reduced most by compaction, so that the influence of compaction is greater at higher potentials.

Residual water content, θ_r, may be thought of as water which cannot be withdrawn from a soil by suction, but in practice is often a fitting parameter. Saturated water content, θ_s, is a measurable quantity which is usually less than soil porosity because of entrapped air.

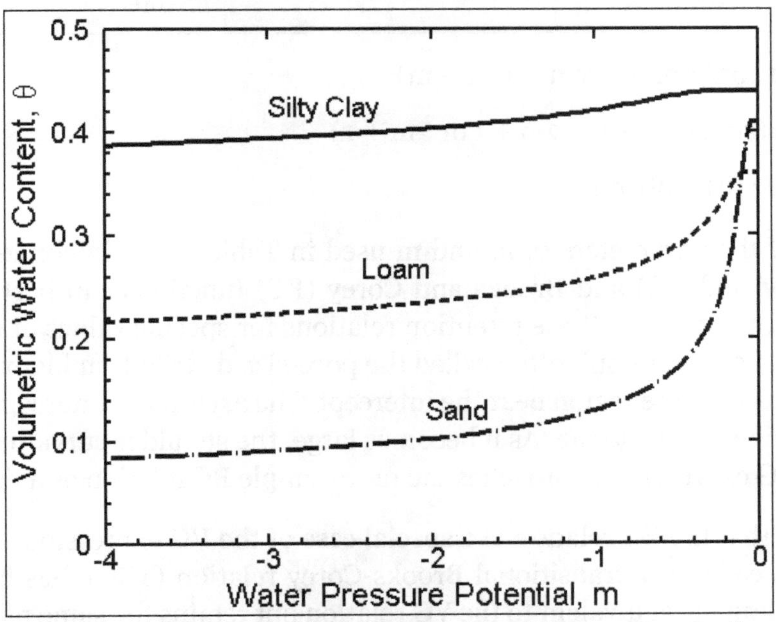

Generalized water retention relations for three different textured soils.

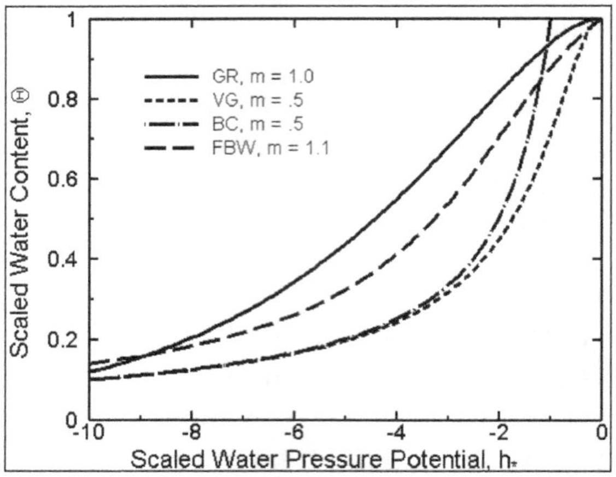

Examples of various algebraic forms for describing the soil water retention relationship.

Table: Functional Relationship for hydraulic Characteristics.

Function (and abbreviation)	Parameter m	kr = K/K	Θ (h*) Relation
Gardner (GR)	m>0	exp(h*)	$\Theta = \left[\exp(h^*/2)(1-h^*/2)\right]^{2/(m+2)}$
van Genuchten[a] (VG)	0<m<1	$\Theta^P\left[1-\left(1-\Theta^{1/m}\right)^m\right]^2$	$\Theta = \left(1+\lvert h_*\rvert^n\right)^{-m}$
Brooks and Corey[b] (BC)	m>0	Θ^v	$\Theta = \lvert h_*\rvert^{-m(1-m)}; h_* > 1$ $= 1; 1 < h_* < 0$
Broadbridge and White (FBW)	m>1	$\dfrac{(m-1)\Theta^2}{(m-\Theta)}$	$h^* = 1-\dfrac{1}{\Theta}+\dfrac{1}{m}\ln\left[\dfrac{(m-1)\Theta}{m-\Theta}\right]$
Linear[c]	None	Θ	$h_* = \ln\Theta$

[a] Use p = 0.5 and commonly use n = 1/(1 – m)

[b] Use v = 2m + 3 (sometimes v = 2m + 1 or 2m + 2)

[c] Also may have kr (h) = dθ/dh

The significance of the parameters α, n, and m used in Table above in connection with the van Genuchten (abbreviated VG) and Brooks and Corey (BC) functions can best be seen in Figure above, which is a log-log plot of these retention relations for specific values of m. The log slope of the asymptote is mn or m/(1 – m), often called the pore-size distribution index, and n determines the degree of curvature in the region near the intercept. The asymptote intercept is 1/α and is often referred to as the air-entry head, he. As n becomes large, the shoulder curvature near he becomes sharper, and the VG expression approaches the more simple BC relation as a limit.

It should be noted that the BC relation is a special case of the VG expression. A generalized form of the BC relation, called the transitional Brooks-Corey relation (TBC), has been introduced by Smith. This is functionally equivalent to the VG relation but retains the same parameters as the BC expression. The relation of m to n often used in the VG expression is not retained.

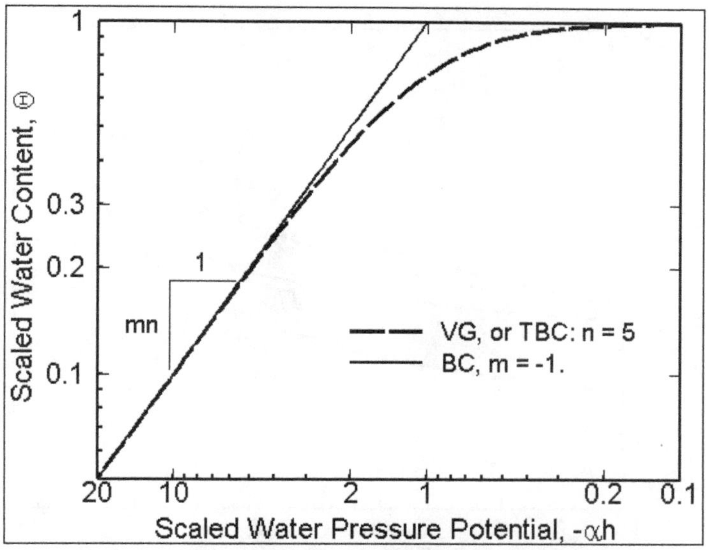

The parameters in the TBC or VG retention relation have specific relationship to the shape of the curve, as illustrated here.

The term soil water capacity, C(h), refers to the slope ($d\theta/dh$) of the soil water characteristic at any point on the curve. This value represents the change in water content per unit change in matric potential and represents an important property for soil water storage and release.

The soil water characteristic can be used to estimate the amount of water "released" between any two potentials. Although the soil water potential largely determines the ease with which a plant can obtain water, it is also important to know how much water is in the soil at potentials above a given critical level. This, along with crop water requirements, allows one to estimate the need for irrigation. The soil water potential will decrease as a plant withdraws water. If hc is considered a critical level below which it is not desired to deplete water, then the amount of water available at a potential h_1, $h_1 > h_c$, will be $\theta(h_1) - \theta(h_c)$. Many soils swell and shrink with wetting and drying, so that all of the water does not come from a constant volume of soil. This is especially important at high potentials where soil structure influences the characteristic.

Hysteresis

The soil water characteristics for sorption and desorption will often differ because the water content in a soil at a given potential depends upon the wetting and drying history of the soil. This history dependence in the relationship between potential and water content is called hysteresis. A schematic example of desorption and sorption curves for a soil is given in Figure below. When the desorption curve is obtained by drying an initially saturated sample and the sorption curve is obtained by wetting an initially dry sample, the two moisture characteristics are known as the primary hysteresis loops or main branches (main drying curve, MD, and wetting curve, MW, respectively). If the soil is not completely dry before rewetting or not completely wet before drying, the resulting curves will fall between the two primary curves, and they are known as scanning curves (wetting scanning curve, WS, and drying scanning curve, DS). Wherever the starting point is within the main curves, a drying condition will approach the MD curve, and a wetting condition will approach the MW curve.

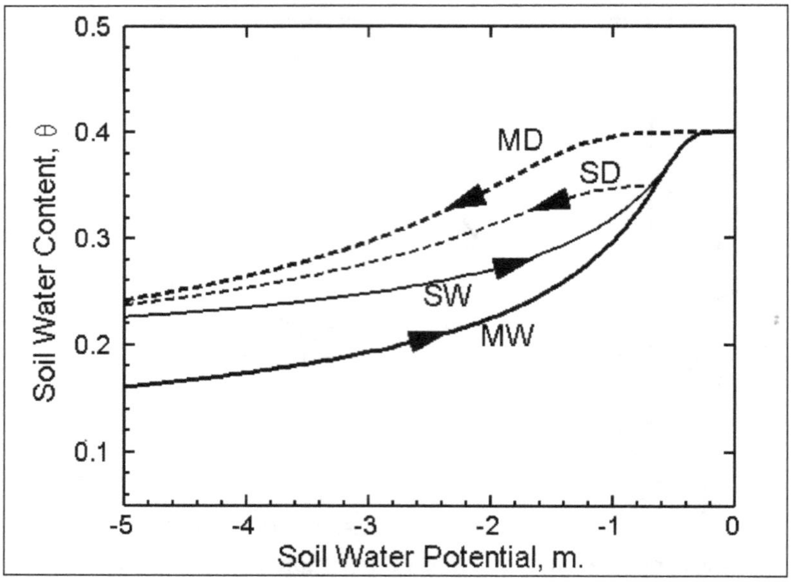

Definition diagram of the hysteresis loops which can occur during wetting and drying of a soil.

At any given potential the water content will be greater in a drying soil (desorption) than in a wetting soil (sorption). Field soil is rarely either completely wet before drying, or completely dry before wetting, so measured primary wetting or drying retention curves can be used only with reservation in interpreting soil water status. The water content or potential can only be estimated from measurement of the water content and the main curves, unless the wetting history is accurately known. However, the amount of error involved is relatively small, compared with other errors involved such as soil variability, climatic changes, and plant variabilities.

Methods of Determining the Soil Water Characteristic

The soil water characteristic is usually determined in the laboratory using tension tables or pressure plates. In all of the techniques used, a porous membrane or plate hydraulically connects the soil water with water in the lower chamber. The pores in the membrane are small enough that, under the imposed pressure, water but not air can pass through. In all cases $P_1 > P_2$, so that water is forced from the soil into the lower chamber. At equilibrium, the imposed pressure (expressed in suitable terms) can be considered as the potential of the water remaining in the soil.

For high potentials the membrane may be blotter paper, fine sand, sintered glass, porous steel, or similar materials. In this case P_1 is often atmospheric and P_2 is obtained with a hanging water column or with regulated vacuum. At lower potentials of about −1 bar or less, the pores of these materials are too large to remain water filled and air will pass through the membrane. A fine-pored ceramic is then used as the membrane, P_2 is atmospheric, and P_1 is obtained with compressed gas, usually air or nitrogen. Ceramic membranes are available with bubbling pressures of 100 bars and more. The air entry value of the plate should be somewhat matched to the soil water potential of interest as the finer-pored ceramics necessary for higher pressures tend to restrict flow. In addition to porous ceramics, porous stainless steel and plastic materials can be appropriate for the wet range.

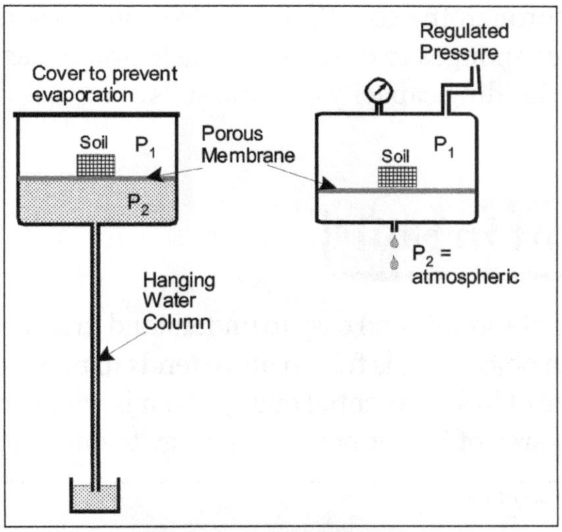

Two methods of determining water retention relations for a soil sample: (left) hanging column and (right) pressure plate.

In practice, a sample of soil is placed in the pressure chamber in a retaining ring and saturated overnight. The desired pressure is then applied until outflow ceases and the soil water is considered to be in equilibrium with the applied pressure. The amount of water in the soil is then determined, usually by oven drying. The process is repeated, with a second sample being subjected to a different pressure. The resulting soil water contents are plotted against applied pressure or vacuum, expressed as potential units (usually cm or millibars) to form the water characteristic. Because the pore size distribution has such a large influence on water retention at high potentials, disturbed samples (dried and sieved) often give erroneous results. Use of so-called undisturbed soil cores is preferable, but even with these, some error is inevitable because of the swelling and shrinking that accompanies wetting and drying of many soils. At low potentials (approaching –15 bars) the soil-specific surface dominates water retention and the error introduced by using disturbed soil samples is quite small.

Soil Moisture

Soil moisture is the water stored in the soil and is affected by precipitation, temperature, soil characteristics, and more. These same factors help determine the type of biome present, and the suitability of land for growing crops. The health of our crops relies upon an adequate supply of moisture and soil nutrients, among other things. As moisture availability declines, the normal function and growth of plants are disrupted, and crop yields are reduced. And, as our climate changes, moisture availability is becoming more variable.

Soil Moisture Activity

Solids, liquids, and gasses, the three phases of matter, are always present in soil. Small mineral and organic particles comprise the solid fraction, and there are spaces (pores) between the solid particles. Some pores are large, and others are very small. Air and water, the gas and liquid phases, exist in the pores. The size of the soil particles and pores affects how much water a soil can hold, and how that water moves through the soil. Sponges also contain solid material and pores of different sizes. This activity uses sponges to observe the basic properties about how soils hold water, and then observes the water holding capacity of some soils.

Water Movement in Soil

Water movement in soils is quite simple and easy to understand in some ways and quite complex and difficult to grasp in others. An object that is free to move tends to move spontaneously from a state of higher potential energy to one of lower potential energy. So it is with water. A unit volume or mass of water tends to move from an area of higher potential energy to one of lower potential energy.

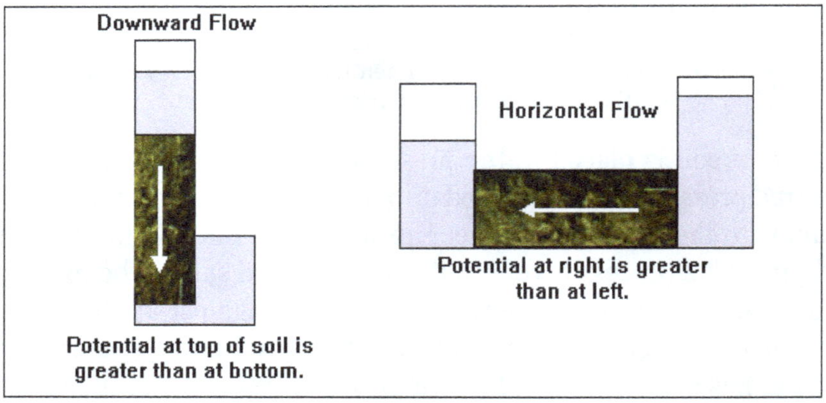

What gives rise to differences in potential energy of a unit of water in a soil? The force of gravity is one factor. Just as water at a higher elevation on a street tends to run down to a lower elevation due to gravity, so water in a soil tends to move downward due gravity. A second factor is the attraction of the soil surfaces for water. When water is added to the bottom of a dry pot of soil, the water moves up into the soil due to this attraction of the soil surfaces for water. The energy level of the water in contact with the soil particles is less than that of the pool of water in the pan so it moves up into the soil. As the soil in the pot becomes wet, this attraction is reduced so that by the time the pores are completely filled, the soil no longer attracts additional water. If a soil is saturated, a third source of potential energy can exist in the form of external pressure such as that provided by a pump or a layer of water in a flooded area. These are the main sources of potential energy in soil water.

Direction of Water Movement

The total potential energy of water is the sum of the potentials from all sources. Potential energy per unit mass or per unit volume or per unit weight is known as the potential of the water. So water free to move will move from a region where it has higher total potential to one of lower total potential. The potential due to gravity is known as the gravitational potential and that due to the soil particles is the matric potential. Soils whose pores are not filled have matric potentials less than zero. Saturated soils under the influence of external hydrostatic pressures have matric potentials which are greater than zero. The total potential at any point is just the sum of the gravitational and matric (or pressure) potentials at that point. The distribution of total potential within a soil allows us to determine if water will move and the direction of movement for any soil system. If the total potentials are equal, no movement will occur.

Rate of Water Movement

Here, we will examine factors that determine the rate of movement. It is useful to think of water movement as the product of a driving force causing water to move and a factor representing the ease with which water moves in the soil. This was formalized by Henry Darcy in 1856 as:

$$q = -K\frac{dTH}{dx}$$

Where q is the volume of water flowing through a unit cross-sectional area of soil per unit time, K is the saturated hydraulic conductivity of the soil, TH is the total hydraulic head and x is the position coordinate in the direction of flow. This equation is known as Darcy's Law. For uniform saturated soils, it is useful to write this equation as:

$$q = K\frac{TH_A - TH_B}{L_{AB}}$$

Where TH_A is the total head at the inlet end of the soil, TH_B is the total head at the outlet end of the soil column, and L_{AB} is the distance between the inlet and outlet. The hydraulic conductivity, K, represents the ease with which water flows through a soil. Its value depends upon the soil properties and the properties of the soil water. The driving force, df, is represented by:

$$df = -\frac{dTH}{dx} \text{ or } \frac{TH_A - TH_B}{L_{AB}}$$

Hydraulic Conductivity

Since some of the volume of a soil is occupied by soil particles, water does not flow through a soil as easily as it does through an open pipe. The ease of water movement is known as the hydraulic conductivity of the soil. Soils with large pores such as sandy soils tend to conduct water more easily than soils with smaller pores such as clay soils. Thus sands have higher saturated conductivity values than clays (Examples). As soils dry out, water in the largest pores is lost first. Water movement must then occur through smaller pores and in films near soil particles. These pores and films present more resistance so the conductivity of a soil decreases as it dries out (Examples). In nature the wetness of the soil often changes with position in the soil. Thus, the conductivity changes with position as well.

Driving Force

The rate of water movement in soils increases with increasing driving force. In saturated soils the driving force is a result in differences in elevation and positive external pressures in the soil. In unsaturated soils, the attraction of the soil surfaces for water is often a major component to the driving force. Driving forces are commonly much larger in unsaturated soils than in saturated soils. The driving force commonly changes in magnitude with changing soil properties and wetness.

Soil-Water Movement

- Saturated Flow
- Unsaturated Flow
- Water Vapor

Movement Saturated Flow

This occurs when the soil pores are completely filled with water. This water moves at water potentials larger than – 33 kPa. Saturated flow is water flow caused by gravity's pull. It begins with infiltration, which is water movement into soil when rain or irrigation water is on the soil surface. When the soil profile is wetted, the movement of more water flowing through the wetted soil is termed percolation.

Hydraulic conductivity can be expressed mathematically as:

$$V = kf$$

Where,

V = Total volume of water moved per unit time.

f = Water moving force.

k = Hydraulic conductivity of soil.

Factors Affecting Movement of Water

- Texture
- Structure
- Amount of organic matter
- Depth of soil to hard pan
- Amount of water in the soil
- Temperature
- Pressure

Vertical Water Flow

The vertical water flow rate through soil is given by Darcy's law. The law states that the rate of flow of liquid or flux through a porous medium is proportional to the hydraulic gradient in the direction of floe of the liquid. (dw) At:

$$QW = -k\,(dw)\,At/Ds$$

Where,

QW = Quantity of water in cm-1

k = rate constant (cm/s)

dw = Water height (head), cm

A = Soil area (cm 2))

t = Time

ds = Soil depth (cm)

Unsaturated Flow

It is flow of water held with water potentials lower than- 1/3 bar. Water will move toward the region of lower potential (towards the greater "pulling" force). In a uniform soil this means that water moves from wetter to drier areas. The water movement may be in any direction .The rate of flow is greater as the water potential gradient (the difference in potential between wet and dry) increases and as the size of water filled pores also increases. The two forces responsible for this movement are the attraction of soil solids for water (adhesion) and capillarity. Under field conditions this movement occurs when the soil macropores (non-capillary) pores with filled with air and the micropores (capillary) pores with water and partly with air. Unsaturated flow is also affected in a similar way to that of saturated flow. Amount of moisture in the soil affects the unsaturated flow. The higher the percentage of water in the moist soil, the greater is the suction gradient and the more rapid is the delivery.

Water Vapor Movement

The movement of water vapor from soils takes place in two ways: (a) Internal movement—the change from the liquid to the vapor state takes place within the soil, that is, in the soil pores and (b) External movement—the phenomenon occurs at the land surface and the resulting vapor is lost to the atmosphere by diffusion and convection.

The movement of water vapor through the diffusion mechanism taken place from one area to other soil area depending on the vapor pressure gradient (moving force).This gradient is simply the difference in vapor pressure of two points a unit distance apart. The greater this difference, the more rapid the diffusion and the greater is the transfer of water vapor during a unit period.

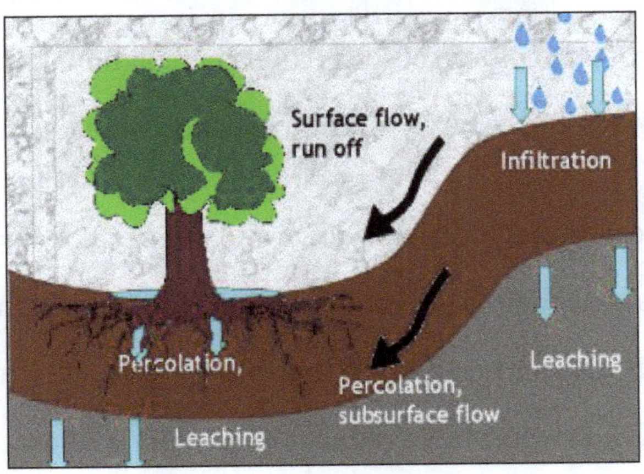

Soil Conditions Affecting Water Vapor Movement

There are mainly two soil conditions that affect the water vapor movement namely moisture regimes and thermal regimes. In addition to these, the various other factors which influence the moisture and thermal regimes of the soil like organic matter, vegetative cover, soil color etc. also affect the movement of water vapor. The movement takes place from moist soil having high vapor pressure to a dry soil (low vapor pressure). Similarly the movement takes place from warmer soil regions to cooler soil region. In dry soils some water movement takes place in the vapor form and such vapor movement has some practical implications in supplying water to drought resistant plants.

Entry of Water into Soil

- Infiltration: Infiltration refers to the downward entry or movement of water into the soil surface.
- It is a surface characteristic and hence primarily influenced by the condition of the surface soil.
- Soil surface with vegetative cover has more infiltration rate than bare soil.
- Warm soils absorb more water than colder ones ϖ Coarse surface texture, granular structure and high organic matter content in surface soil, all help to increase infiltration.
- Infiltration rate is comparatively lower in wet soils than dry soils.

Factors Affecting Infiltration

- Clay minerals
- Soil Texture
- Soil structure
- Moisture content
- Vegetative cover
- Topography

Percolation

The movement of water through a column of soil is called percolation. It is important for two reasons:

- This is the only source of recharge of ground water which can be used through wells for irrigation.
- Percolating waters carry plant nutrients down and often out of reach of plant roots (leaching).
    - In dry region it is negligible and under high rainfall it is high.
    - Sandy soils have greater percolation than clayey soil.
    - Vegetation and high water table reduce the percolation loss.

Permeability

It indicates the relative ease of movement of water with in the soil. The characteristics that determine how fast air and water move through the soil is known as permeability. The term hydraulic conductivity is also used which refers to the readiness with which a soil transmits fluids through it.

Drainage

The frequency and duration of periods when the soil is free from saturation with water. It controls the soil cum water relationship and the supply of nutrients to the plants.

Drainage Class

- Very poorly drained
- Poorly drained
- Imperfect
- Moderately well
- Well
- Somewhat excessive
- Excessive
- Hysteresis

The moisture content at different tensions during wetting of soil varies from the moisture content at same tensions during drying. This effect is called as hysteresis. This is due to the presence of capillary and non-capillary pores. The moisture content is always low during sorption and high during desorption. Hysteresis phenomenon exists in soil minerals as a consequence of shrinking and swelling. Shrinking and swelling affect pore size on a micro basis as well as on the basis of overall bulk density. So, hysteresis phenomenon occurs due to factors like shape and size of soil pores and their interconnection with each other pore configuration, nature of soil colloids bulk density of soil and entrapped air. The most important factor affecting hysteresis is the entrapment of air in the soil under rewetting condition. This clogs some pores and prevents effective contact between others.

Measuring Soil-Water Content

To measure soil-water content change for the purpose of scheduling irrigation, monitoring should be done at several locations in each field and at different soil depths (6" increments). Most devices used indicate relative soil-water values and are difficult to calibrate to relate to specific quantitative values. A calibration curve for each specific kind of soil and soil water content (tension) should be available with the device or needs to be developed. Methods and devices to measure or estimate soil-water content include:

Soil Feel and Appearance Method: Soil samples are collected at desired depths and compared to tables or pictures that give moisture characteristics of different soil textures. Refer to NRCS color publication, Estimating Soil Moisture by Feel and Appearance. How soil samples taken in the field from appropriate locations and depths feel and look gives some indication of moisture content. A shovel can be used to get samples, but for some soils a soil auger or a sampling tube is better. The appearance and feel of a handful of soil that has been squeezed very firmly can be compared with descriptions in a guide of estimated available-moisture content for different soil textures and conditions. Table below is a guide that has been used for some time. The feel and appearance method is one of the cheapest and easiest methods to use for estimating water content, but it does require some work to get soil samples. Although this method is not the most accurate, with experience and judgment the irrigator should be able to estimate the moisture level within a reasonable degree of accuracy.

Gravimetric or Oven Dry Method: Soil samples are collected using a core sampler. Samples must be protected from drying before they are weighed. Samples are taken to the office work room, weighed (wet weight), oven dried, and weighed again (dry weight). An electric oven takes 24 hours at 105 degrees Celsius to adequately remove soil water. Percentage of total soil-water content on a dry weight basis is computed. To convert to a volumetric basis, the percentage water content is multiplied by the soil bulk density. Available soil water is calculated by subtracting percent total soil water at wilting point. This procedure is the most accurate method for determining the soil water content, but is time consuming and may be impractical for most farmers.

Table: Practical interpretation chart on soil moisture for soil textures and conditions.

Feel or appearance of soil				
Available moisture in soil	Coarse-textured Soils	Moderately Coarse Textured soils	Medium textured Soils	Fine and very Fine textured Soils
0 percent	Dry, loose, and single grained; flows through fingers	Dry and loose; flows through fingers	Powdery dry; in some places slightly crusted but breaks down easily into powder	Hard, baked, and cracked; has loose crumbs on surface in some places
50 percent or less	Appears to be dry; does not form a ball under pressure	Appears to dry; does not form a ball under pressure	Somewhat crumbly but holds together under pressure	Somewhat pliable; balls under pressure
50 to 75 percent	Appears to be dry; does not form a ball under pressure	Balls under pressure but seldom holds together	Forms a ball under pressure; somewhat plastic; slicks slightly under pressure	Forms a ball; ribbons out between thumb and forefingers

75 percent to field capacity	Sticks together slightly; may form a very weak ball under pressure	Forms weak ball that breaks easily; does not slick	Forms ball; very pliable; slicks readily if relatively high in Clay	Ribbons out between fingers easily; have a slick feeling
At field capacity(100 percent)	On squeezing, no free water appears on soils but wet outline of ball is left on hand	Same as for coarse textured soils at field capacity	Same as for coarse- textured soils at field capacity	Same as for coarse textured soils at field capacity
Above field capacity	Free water appears when soil is bounced in hand	Free water is released with kneading	Free water can be squeezed out	Puddles; free water forms on surface

1/ Ball are formed by squeezing a handful of soil very firmly.

The following equipment is required:

- Moisture proof seamless aluminum or tin sample boxes (cans) with a capacity of 3 ounces or more to contain sample for drying.
- Beam balance with a minimum of 500-gram capacity, accurate to 0.1 gram.
- Drying oven, with thermometer, capable of maintaining a temperature of 2200 F to 2400F. (105°C to 115°C).
- A core soil sampler to take bulk density samples. Alternatively, the bulk density can be determined by the sand cone method or water balloon method.

Procedure

- Collect a representative soil sample of known volume using a core sampler.
- Determine the wet weight of the sample (WW).
- Dry the sample in the oven at 1050 C to 1150 C until it attains a constant weight. This will take about 24 hours. Check the weight at one hour intervals near the end of the 24 hour period. When there is no weight change, the sample is dry.
- Determine the dry weight of the sample (DW). Remember to deduct the tare weight of the container when determining wet weight and dry weight.
- Compute the bulk density (BD) and total soil water content (TSWC):

$$BD = \frac{DW(g)}{\text{Sample Volume}(cc)}(\text{dry weight basis})$$

Weight of water lost.

$$WAT = WW - DW$$

Percent Water Content (dry weight basis),

$$WC = \frac{WAT}{DW} \times 100$$

$$TSWC = \frac{BD \times WC}{100} (\text{inches water per inch soil depth})$$

- The total soil water content (TSWC) includes moisture that is not available to the plant at the permanent wilting point. The permanent wilting point (Pw), the point at which a plant can no longer obtain enough soil water to meet transpiration needs, occurs at 15 atmospheres of tension and is normally determined in the laboratory. When laboratory data are not available, one of the following procedures can be used to estimate the wilting point.
    - This procedure requires knowledge of the percent clay, less any clay size carbonate particles, of the soil being measured. For many soils, this procedure will provide a close estimate of the wilting point. The wilting point is calculated with use of the following equation:

Pw = 0.4 x clay (%)

Where, Pw is the wilting point, using the clay content, expressed as percent of water on a dry weight basis.

- This procedure can be used where only the soil texture is known. The values given represent and average wilting point for the given texture. Pw is expressed as percent water on a dry weight basis. The values given do not apply to soils having soil fragments larger than 2.0 millimeters.

Texture	Pw
Clay	25
Silty clay	19
Sandy clay	17
Silty clay loam	13
Clay loam	13
Sandy clay loam	11
Silt loam	5.5
Loam	7
Very fine sandy loam	4
Fine sandy loam	4
Sandy loam	4
Loamy fine sand	3
Loamy sand	3
Fine sand	2
Sand	2

- To determine the soil water content (SWC), Pw percentage is first converted to inches of water (WP). WP is then subtracted from TSWC to obtain SWC.

$$WP = \frac{BD \times Pw}{100} \text{ (inches water per inch soil depth)}$$

SWC = TSWC-WP (inches/inch soil depth)

- To determine the SWC for a given increment of depth, multiply SWC by the depth, in inches, being evaluated.

These measurements are made after an irrigation or effective rainfall. Before making the measurements, allow about 24 hours for sand and about 48 hours for clay for the gravitational water to drain. Determine AWC using the above procedure, where SWC becomes AWC.

Chemical Drying Method (Speedy Moisture Tester)

This method of drying soil samples is based on the principle that a given quantity of moisture, when combined with calcium carbide, will react to produce a specific volume of gas (acetylene). By applying this principle, a device known as the Speedy Moisture Tester was developed in England and is commercially available. It confines in a cylindrical pressure chamber the gas produced from this reaction. One end of the pressure chamber is equipped with a cap for inserting the carbide and soil sample and a clamping arrangement to confine the gas during the test. The gas pressure is read on a gage located on the other end of the pressure chamber. The gage is calibrated for a 26-gram wet weight sample of soil and the reading can be converted readily to a dry weight moisture percentage by use of a calibration curve supplied with the instrument.

The calcium carbide gas pressure method is a quick and reasonably accurate way of determining the moisture content of a soil. The time required to dry a sample is about 3 minutes. Its simplicity of use is such that anyone can become proficient enough in a short time to make its use a routine operation.

Care must be taken in measuring the 26- gram soil sample. When the test is finished, examine the sample for lumps. If the soil was not completely broken down by the steel balls, retest and increase the shake-and-rest time by one minute. The calcium carbide gas pressure method should not be used on saturated highly organic soils.

Tensiometers

Soil-water potential (tension) is a measure of the amount of energy with which water is held in the soil. Tensiometers work on the principle that a partial vacuum is created in a closed chamber when water moves out through a porous ceramic tip to the surrounding soil. Tension is measured by a water manometer, a mercury manometer, or a vacuum gage. The scales are generally calibrated in either hundredths of an atmosphere or in centimeters of water. Tensiometers that utilize a mercury manometer are usually preferred as research tools because they afford great precision. Because of their simplicity, tensiometers equipped with Bourdon vacuum gages are better suited to practical use and to irrigation control.

After the cup is placed in the soil at the desired depth, the instrument must be filled with water. Water moves through the porous cup until the water in the cup and the water in the soil reach equilibrium.

Any increase in tension that occurs as the soil dries causes the vacuum-gage reading, which can be read above ground, to increase. Conversely, an increase in soil-water content reduces tension and lowers the gage reading. The tensiometer continues to record fluctuations in soil-water content unless the tension exceeds 0.85atmosphere, at which point air enters the system and the instrument ceases to function. Then after an irrigation or rain, the instrument must again be filled with water before it can operate.

Some experience is required to use a tensiometer. If air enters the unit through any leaks at the rubber connections, measurements are not reliable. Air leaks can result from faulty cups. They may occur also at the contact points of the setscrews used to secure the porous cup to the metal support. Some manufacturers provide a test pump that can be used to test the gage and to remove air from the instrument.

Tensiometers readings reflect soil-moisture tension only; that is, they indicate the relative wetness of the soil surrounding the porous tip. They do not provide direct information on the amount of water held in the soil. Tension measurements are useful in deciding when to irrigate, but they do not indicate how much water should be applied. A special moisture characteristic curve for the particular soil is needed to convert moisture-tension measurements into available-moisture percentages.

Table: Guidelines for using soil moisture tension data to schedule irrigation event.

	25% MAD (cb)	50% MAD (cb)
Coarse Sand	12	20
Sand	12	20
Fine Sand	12	20
Loamy Sand	15	25
Loamy Coarse Sand	15	25
Loamy Fine Sand	15	25
Loamy Very Fine Sand	20	40
Sandy Loam	20	40
Fine Sandy Loam	25	50
Very Fine Loam	25	50
Loam	30	60
Silt Loam	40	85
Sandy Clay Loam	40	85
Clay Loam	45	90
Silty Clay Loam	45	90
Sandy Clay	50	95
Silty Clay	50	95
Clay	50	95

cb = Centi bars*

Tensiometers do not satisfactorily measure the entire range of available moisture in all soil types. But they probably are the best field instruments to use to determine moisture conditions in the wet range. They are best suited to use in sandy soils, since in these soils a large part of the moisture available to plants is held at a tension of less than I atmosphere. Tensiometers are less well suited to use in fine-textured soils, which hold only a small part of the available moisture at a tension of less than 1atmosphere.

Tensiometers installed at different rooting depths have different gauge readings because of soil water potential change in rooting depths. With uniform deep soil, about 70 – 80% of soil moisture withdrawal by plant roots is in the upper half of the rooting depth. Recommended depths for setting tensiometers are given in Table below.

Plant root zone depth (in)	Shallow tensiometer (in)	Deep Tensiometer (in)
18	8	12
24	12	18
36	12	24
>48	18	36

Installing tensiometers must be done carefully and good maintenance is required for accurate and reliable results. They also must be protected against freeze damage.

Maintenance kits that include a hand vacuum pump are required for servicing tensiometers. The hand pump is used to draw out air bubbles from the tensiometer and provide equilibrium in tension. Tensiometers should be installed in pairs at each site, at one-third and two-thirds of the crop rooting depth. A small diameter auger (or 1/2" steel water pipe) is required for making a hole to insert the tensiometer. Figure below shows a tensiometer and gauge and illustrates installation and vacuum pump servicing.

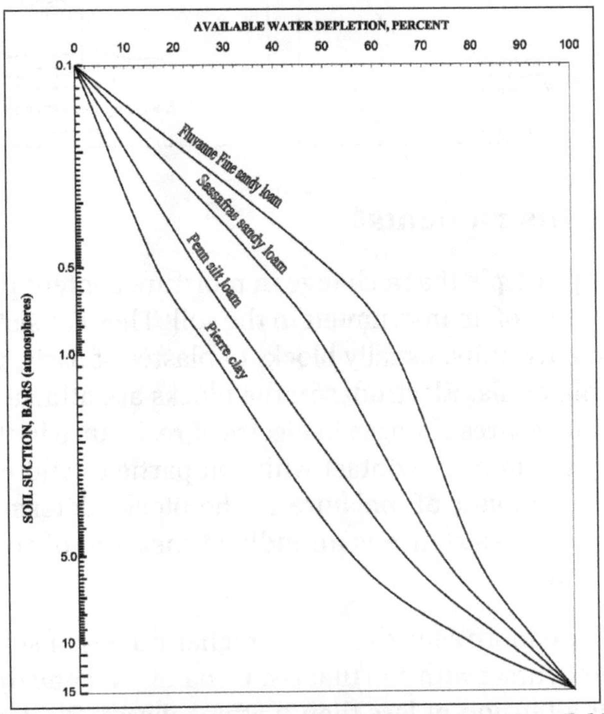

Water retention curves for several soils plotted in terms of percent available water removed.

Tensiometer's Installation, Gauge and Servicing

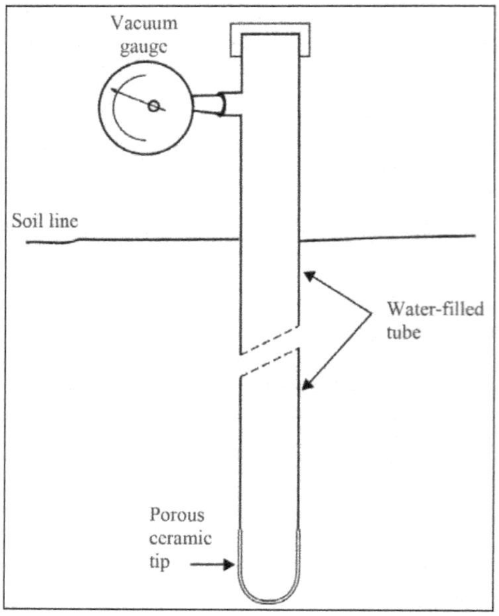

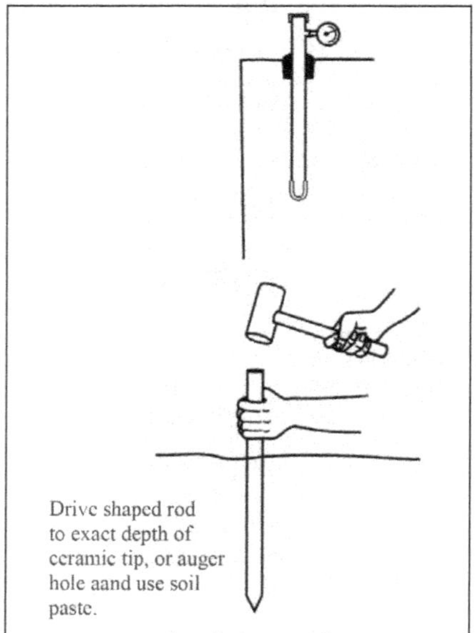

Installation procedure

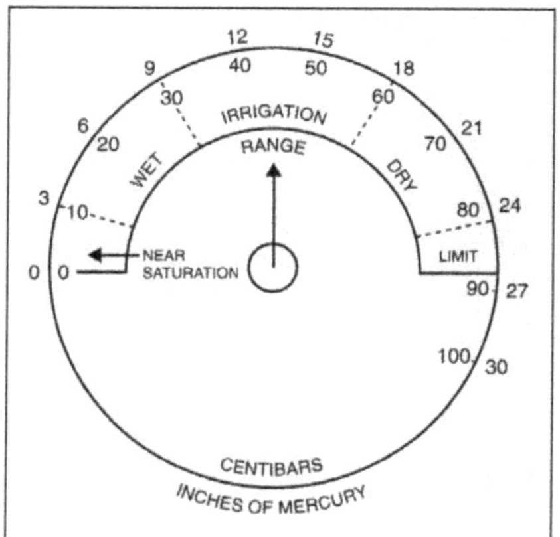

Vacuum guage

Servicing tensiometer using a vacuum pump

Electrical-Resistance Instruments

These instruments use the principle that a change in moisture content produces a change in some electrical property of the soil or of an instrument in the soil. They consist of two electrodes permanently mounted in conductivity units, usually blocks of plaster of paris, nylon, fiberglass, gypsum, or combinations of these materials. Electrodes in the blocks are attached by wires to a resistance or conductance meter that measures changes in electrical resistance in the blocks. When the units are buried in the soil, they are in close contact with soil particles and respond to changes in soil moisture content. Since the amount of moisture in the blocks determines electrical resistance, measurement of any change in resistance is an indirect measure of soil moisture if the block is calibrated for a particular soil.

Nylon and fiberglass units are more sensitive in the higher ranges of soil moisture than plaster of Paris blocks, but often their contact with soil that is alternately wet and dry. is not very good. Nylon units are most sensitive at a tension of less than 2 atmospheres. Plaster of Paris blocks function most effectively at a tension between 1 and 15 atmospheres, and fiberglass units operate satisfactorily over the entire range of available moisture. A combination of fiber glass and plaster of Paris provides sensitivity in both the wet and dry range and provides good contact between the soil and the unit.

Electrical-resistance instruments are sensitive to salts in the soil; fiberglass units are more sensitive than plaster of Paris. Their readings are also affected by concentrations of fertilizer. Where fertilizer is spread in bands, the unit should be placed well to one side of the bands. Temperature also affects reading in all units, but much less than other sources of variation. In some units, calibration drift has caused changes of as much as I atmosphere of tension in a single season. The magnitude of a change depends on the number of drying intervals and the number of days between each. Readings also vary with soil type. Since the same reading may indicate different amounts of available moisture for different soil textures, the instrument must be calibrated for the soil in which it is to use. For good accuracy, each instrument site shall be calibrated. Due to the possibility of calibration drift, particularly if fertilizer is applied with irrigation, the calibration should be checked once or twice each growing season. Calibration should be done with an accurate method, such as the gravimetric or the chemical drying methods.

If readings are to be representative of an area, the blocks must be properly installed. Individual blocks must be placed in a hole, which disturbs the soil. If the soil is not replaced in the hole at the same density and in the same way as in the rest of the profile, the root-development and moisture pattern may not be representative. A good method is to force the block into undisturbed soil along the sides of the hole dug for placement of the blocks. In one type, the blocks are cast in a tapered stake. A tapered hole, the same size as the stake, is bored into the ground with a special auger. The stake is saturated with water and then pushed into the hole so that close contact is made between the stake and the soil.

Most of the commercial instruments give good indications of moisture content if they are used according to the manufacturer's instructions. For good results, however, the blocks need to be calibrated in the field for each job. Experience and careful interpretation of instrument readings are needed to get a good estimate of soil moisture conditions. Electrical conductivity sensors are becoming a popular tool, and are recommended on finer textured soils with a high water holding capacity. These instruments are calibrated to read centibars of soil water tension correlated to tensiometer readings. Since they are actually measuring electrical conductivity which is converted to centibars of tension, they can operate in a much higher tension range resulting in more accurate readings. However in sandy soils they are not as sensitive as the tensiometer with a longer response time to soil moisture changes.

Table: Interpretations of readings on typical electrical resistance meter.

Soil water condition	Meter readings 1/ (0—200 scale)	Interpretation
Nearly saturated	180 — 200	Near saturated soil often occurs for a few hours following irrigation. Danger of water logged soils, a high water table, or poor soil aeration if readings persist for several days.
Field capacity	170 — 180	Excess water has mostly drained out. No need to irrigate. Any irrigation would move nutrients below irrigation depth (root zone).

Irrigation range	80 — 120	Usual range for starting irrigations. Soil aeration is assured in this range. Starting irrigations in this range generally ensures maintaining readily available soil water at all times.
Dry	<80	This is the stress range; however, crop may not be necessarily damaged or yield reduced. Some soil water is available for plant use, but is getting dangerously low.

1/: Indicative of soil-water condition where the block is located. Judgment should be used to correlate these readings to general crop conditions throughout the field. It should be noted, the more sites measured, the more area represented by the measurements.

Dielectric Constant Method

The dielectric constant of material is a measure of the capacity of a non-conducting material to transmit high frequency electromagnetic waves or pulses. The dielectric constant of a dry soil is between 2 and 5. The dielectric constant of water is 80 at frequency range of 30 MHz — 1 GHz. Relatively small changes in the quantity of free water in the soil have large effects on the electromagnetic properties of the soil-water media. Two approaches developed for measuring the dielectric constant of the soil water media (water content by volume) are time domain reflectometry (TDR) and frequency domain reflectometry (FDR).

For TDR technology used in measuring soil water content, the device propagates a high frequency transverse electromagnetic wave along a cable attached to parallel conducting probes inserted into the soil. A TDR soil measurement system measures the average volumetric soil-water percentage along the length of a wave guide. Wave guides (parallel pair) must be carefully installed in the soil with complete soil contact along their entire length, and the guides must remain parallel. Minimum soil disturbance is required when inserting probes. This is difficult when using the device as a portable device. The device must be properly installed and calibrated. Differing soil texture, bulk density, and salinity do not appear to affect the dielectric constant.

FDR approaches to measurement of soil-water content are also known as radio frequency (RF) capacitance technique. This technique actually measures soil capacitance. A pair of electrodes is inserted into the soil. The soil acts as the dielectric completing a capacitance circuit, which is part of a feedback loop of a high frequency transistor oscillator. The soil capacitance is related to the dielectric constant by the geometry of the electric field established around the electrodes. Changes in soil-water content cause a shift in frequency. University and ARS comparison tests have indicated that, as soil salinity increases, sensor moisture values were positively skewed, which suggests readings were wetter than actual condition.

FDR devices commercially available include:

- Portable hand-push probes: These probes allow rapid, easy, but only qualitative readings of soil-water content. Probe use is difficult in drier soil of any texture, soils with coarse fragments, or soils with hardpans. A pilot hole may need to be made using an auger. The probe provides an analog, color-coded dial gauge (for three soil types—sand, loam, and clay), or a digital readout. The volume of soil measured is relatively small (a cylinder 4 inches tall by 1 inch in diameter). Several sites in a field should be measured, and can be, because probes are rapid and easy to use. Proper soil/probe tip contact is essential for accurate and consistent readings.

- Portable device that uses an access tube similar to a neutron gauge: The probe suspended on a cable is centered in an access tube at predetermined depths where the natural resonant frequency or frequency shift between the emitted and received frequency is measured by the probe. The standard access tube is 2-inch diameter schedule 40 PVC pipe. Installation of the access tube requires extreme care to ensure a snug fit between the tube and the surrounding soil. Air gaps or soil cracks between the tube and soil induce error.
- The device is calibrated by the manufacturer to sand and to an average bulk density for sand. Recalibration is required for any other soil texture and differing bulk density. The volume of soil measured is not texture or water content dependent, and approximates a cylinder 4 inches tall and 10 inches in diameter. Accuracy can be good in some soils with proper installation and calibration, and there are no radioactive hazards to personnel such as when using a neutron gauge. Proper installation of the access tube is essential and can be quite time consuming. Accuracy of data is largely dependent on having a tight, complete contact between the access tube and the surrounding soil. Before making a large investment in equipment, it is highly recommended that adequate research be done on comparison evaluations that are in process by various universities and the ARS. Good sources of information are technical papers and proceedings of ASAE, ASCE, and Soil Science Society of America, as well as direct discussion with personnel doing evaluations.
- Other electronic sensors: Numerous sensors are commercially available using microelectronics. Inexpensive devices sold at flower and garden shops measure the electrical voltage generated when two dissimilar metals incorporated into the tip are placed in an electrolyte solution; i.e., the soil water. Most of these devices are sensitive to salt content in the soil-water solution.

Factors to be evaluated for the selection and application of a soil-water content measuring program include:

- Initial cost of device, appurtenances, special tools, and training.
- Irrigation decision maker's skill, personal interest, and labor availability.
- Field site setup, ease of use and technical skill requirements.
- Repeatable readings and calibration requirement.
- Interpretations of readings— qualitative and quantitative needs.
- Accuracy desired and accuracy of device.
- Operation and maintenance costs.
- Special considerations including licensing from NRC (private individuals do not operate under ARS licensing), storage, handling, film badge use, training required, disposal of radioactive devices, and special tools required for access tube installation.

References

- Principles-soil-plant-water-relations: jliedu.com, Retrieved 05, June 2020
- Soil-water-relationships-43265020: researchgate.net, Retrieved 13, April 2020
- Introduction-to-Soil-Science: ecoursesonline.iasri.res.in, Retrieved 09, August 2020
- Soil-profile: soil-net.com, Retrieved 03, March 2020
- Soil-textures: soilsensor.com, Retrieved 17, Jan 2020

Chapter 3

Plant and Water Relations

The potential energy of water in a system compared to pure water under same temperature and pressure conditions is known as water potential. Osmosis is the movement of solvent molecules from a region of higher concentration to the region of lower concentration through a semi-permeable membrane. The chapter closely examines the key concepts of plant and water relations to provide an extensive understanding of the subject.

Knowledge of plant water relations is important because water is essential for both plants and animals. It serves as a medium for the dissolution of substances. A huge amount of water is taken up daily by plants and a considerable amount is lost in transpiration. The water requirement of different categories of plants is different.

Water molecules possess a certain amount of kinetic energy. The greater the concentration of water in a system, the greater is its kinetic energy or water potential. If two systems containing water are in contact, movement of water molecules occurs from the system with higher energy to the system with lower energy. Water potential is expressed in Pascal. The value of water potential of pure water at standard temperature is zero.

If a certain amount of solute is added to pure water, the concentration of water decreases and thus the water potential decreases. The amount by which the water potential decreases is called solute potential. This is always negative and the value of solute potential decreases with an increase in the amount of the dissolved solutes. The value of water potential increases when a pressure more than atmospheric pressure is applied to pure water. When water enters a plant cell by diffusion and exerts a pressure on the walls of the cell, the cell is termed as turgid. This increases the pressure potential. This value is usually positive. Water potential is the sum of solute potential and pressure potential.

Osmosis

Osmosis is a type of diffusion that, in biology, is usually related to cells. Diffusion is when molecules or atoms move from an area of high concentration to an area of low concentration. Osmosis is when a substance crosses a semipermeable membrane in order to balance the concentrations of another substance. In biology, this is usually when a solvent such as water flows into or out of a cell depending on the concentration of a solute such as salt. Osmosis happens spontaneously and without any energy on the part of the cell.

Solvents and Solutes

Osmosis deals with chemical solutions. Solutions have two parts, a solvent and a solute. When solute dissolves in a solvent, the end product is called a solution. Salt water is an example of a solution; salt is the solute, and water is the solvent.

Types of Solutions

In biology, there are three different types of solutions that cells can be in: isotonic, hypotonic, and hypertonic. Different types of solutions have different impacts on cells due to osmosis.

- Isotonic: An isotonic solution has the same concentration of solutes both inside and outside the cell. For example, a cell with the same concentration of salt inside it as in the surrounding water/fluid would be said to be in an isotonic solution. Under these conditions, there is no net movement of solvent; in this case, the amount of water entering and exiting the cell's membrane is equal.
- Hypotonic: In a hypotonic solution, there is a higher concentration of solutes inside the cell than outside the cell. When this occurs, more solvent will enter the cell than leave it to balance out the concentration of solute.
- Hypertonic: A hypertonic solution is the opposite of a hypotonic solution; there is more solute outside the cell than inside it. In this type of solution, more solvent will exit the cell than enter it in order to lower the concentration of solute outside the cell.

How Osmosis Affects Cells

Osmosis affects plant and animal cells differently because plant and animal cells can tolerate different concentrations of water. In a hypotonic solution, an animal cell will fill with too much water and lyse, or burst open. However, plant cells need more water than animal cells, and will not burst in a hypotonic solution due to their thick cell walls; hypotonic solutions are ideal for plant cells. The optimal condition for an animal cell is to be in an isotonic solution, with an equal amount of water and solutes both inside and outside. When a plant cell is in an isotonic solution, its cells are no longer turgid and full of water, and the leaves of the plant will droop. In a hypertonic solution, water will rush out of both animal and plant cells, and the cells will shrivel (in plants, this is called plasmolyzation). This is why slugs and snails shrivel and die when salt is sprinkled onto them; water leaves their cells in order to balance the higher concentration of salt outside the cells.

This figure shows the effects of osmosis on red blood cells.

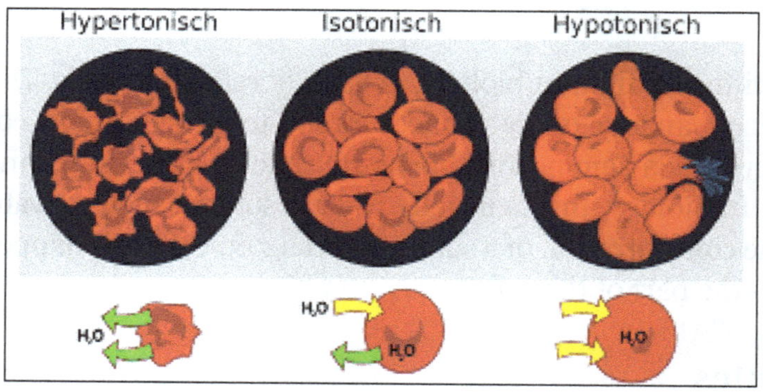

Examples of Osmosis

Osmosis is how plants are able to absorb water from soil. The roots of the plant have a higher solute concentration than the surrounding soil, so water flows into the roots. In plants, guard cells are also affected by osmosis. These are cells on the underside of leaves that open and close to allow gas exchange. When the plant's cells are full of water, the guard cells swell and open the stomata, small holes that allow the plant to take in carbon dioxide and release oxygen.

Osmosis can have adverse effects on animals such as fish. If freshwater or saltwater fish are put into water that has a different salt concentration than they are used to, they will die from having too much water enter or leave their cells. Osmosis can affect humans as well; in a person infected with cholera, bacteria overpopulate the intestines, leaving the intestines unable to absorb water. The bacteria actually reverse the flow of absorption because osmosis causes water to flow out of the intestinal cells instead of in. This causes severe dehydration and sometimes death.

Endosmosis

Endosmosis is the osmosis toward the inside of a cell or vessel. It occurs when the water potential of the cell surrounding is higher than inside the cells. Thus, the solute concentration of the surrounding solution is less than that inside the cytoplasm. This type of solutions is called hypotonic solutions. Water molecules move into the cell across the cell membrane in endosmosis. Cells swell as a result of the movement of water into them.

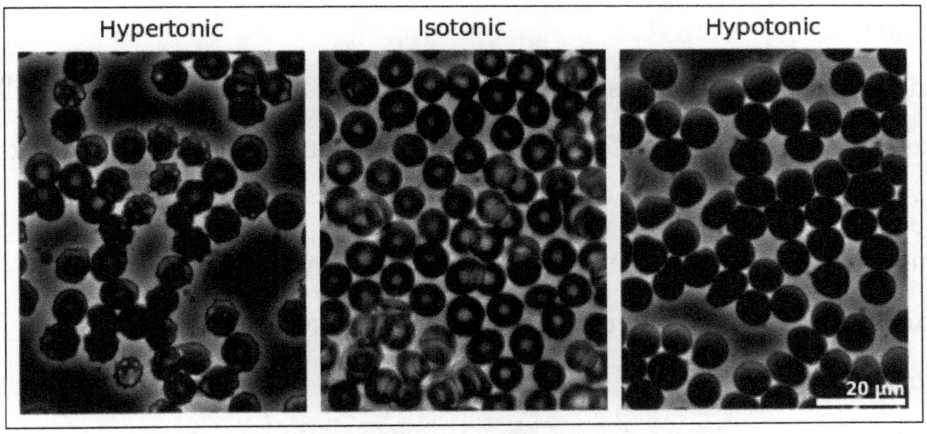

Cells in Hypertonic, Isotonic, and Hypotonic Solutions.

In plants, the absorption of capillary water from the soil by roots and the entrance of water into the xylem vessels occur by endosmosis.

Exosmosis

Exosmosis refers to the osmosis toward the outside of a cell or vessel. It occurs when the water potential of the cell surrounding is lesser than inside the cells. Consequently, the solute concentration of the surrounding solution is higher than that inside the cytoplasm. These types of solutions are called hypertonic solutions. Water molecules move out of the cell across the cell membrane in exosmosis. Cells shrink as a result of the movement of water out of them. Plant cells in the three types of solutions are shown in figure below.

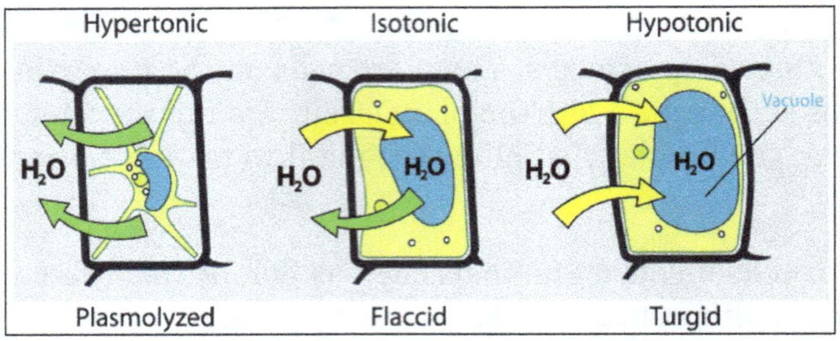

Plant Cells in Hypertonic, Isotonic, and Hypotonic Solutions.

If a cell is placed in a strong hypertonic solution, the cell can be dehydrated and die. This situation is referred to as plasmolysis. In isotonic solutions, cells become flaccid. In hypotonic solutions, cells become turgid. If a cell is placed in a strong hypotonic solution, it may burst. The movement of water from the root hair cells to the cortical cells of the root occurs by exosmosis.

Similarities Between Endosmosis and Exosmosis

- Both endosmosis and exosmosis are two types of osmosis.
- Water molecules move across the cell membrane in both endosmosis and exosmosis.

Difference Between Endosmosis and Exosmosis

- Definition:
  - Endosmosis: Endosmosis refers to the osmosis toward the inside of a cell or vessel.
  - Exosmosis: Exosmosis refers to the osmosis toward the outside of a cell or vessel.
- Water Movement:
  - Endosmosis: Water moves into the cell during endosmosis.
  - Exosmosis: Water moves out of the cell during exosmosis.
- Type of Solutions:
  - Endosmosis: Endosmosis occurs when cells are placed in hypotonic solutions.
  - Exosmosis: Exosmosis occurs when cells are placed in hypertonic solutions.
- Solute Concentration in the Surroundings:
  - Endosmosis: Endosmosis occurs when the solute concentration of the surroundings is less than the solute concentration inside the cell.
  - Exosmosis: Exosmosis occurs when the solute concentration of the surroundings is higher than the solute concentration inside the cell.
- Water Potential:
  - Endosmosis: The water potential of the surroundings is higher than that of the cytosol in endosmosis.

  - Exosmosis: The water potential of the surroundings is less than that of the cytosol in exosmosis.
- Result:
  - Endosmosis: The cells may swell as a result of endosmosis.
  - Exosmosis: The cells shrink as a result of exosmosis.
- Examples:
  - Endosmosis: The absorption of capillary water from the soil by roots and the entrance of water into the xylem vessels are the examples of endosmosis in plants.
  - Exosmosis: The movement of water from the root hair cells to the cortical cells of the root is an example of exosmosis.

Plasmolysis

Plasmolysis is defined as the process of contraction or shrinkage of the protoplasm of a plant cell and is caused due to the loss of water in the cell. Plasmolysis is an example of the results of osmosis and rarely occurs in nature.

The word Plasmolysis was generally derived from a Latin and Greek word plasma – The mould and lusis meaning loosening.

Stages of Plasmolysis

The complete process of Plasmolysis take place in three different stages:

- Incipient plasmolysis: It is the initial stage of the plasmolysis, during which, water starts flowing out of the cell; initially, the cell shrinks in volume and cell wall become detectable.
- Evident plasmolysis: It is the next stage of the plasmolysis, during which, the cell wall has reached its limit of contraction and cytoplasm gets detached from the cell wall attaining the spherical shape.
- Final plasmolysis: It is the third and the final stage of the plasmolysis, during which the cytoplasm will be completely free from the cell wall and remains in the centre of the cell.

Water Transport Across Cell Membrane

During the process of Plasmolysis within the plant cell, the cell membrane separates the interiors of the cell from the surrounding. It allows the movement of water molecules, ion and other selective particles across the membrane and stops others. Water molecules travel in and out of the cell across the cell membranes and the water flow is a necessary consequence that enables cells to fetch water.

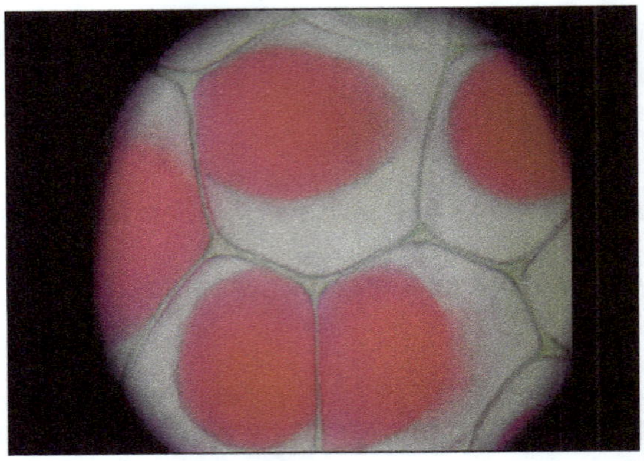

The process of plasmolysis can be easily explained in the laboratory by placing a living cell in a strong salt solution. When the plant cells are placed in the concentrated salt solution, because of osmosis, water from the cell sap moves out. Therefore, the water travels through the cell membrane into the neighboring medium. Finally, the protoplasm separates from the cell and assumes a spherical shape.

Normally, for this experiment, Tradescantia or Rheo plant cell, Elodea plants or onion epidermal cells are used, because they have colored sap which can be easily observed and identified under the microscope.

Types of Plasmolysis

There are two different types of plasmolysis and this classification is mainly based on the final structure of the cytoplasm.

- Concave Plasmolysis: During the concave plasmolysis, both the cell membrane and protoplasm shrink away and begins to detach from the cell wall, which is caused due to the loss of water. Concave plasmolysis is a reversible process and it can be revised by placing the cell in a hypotonic solution, which helps calls to regain the water back into the cell.
- Convex Plasmolysis: During the convex plasmolysis, both the cell membrane and protoplasm lose so much water that they completely get detach from the cell wall. Later, the cell wall collapses and results in the destruction of the cell. Similar to concave plasmolysis, convex plasmolysis cannot be reversed, and this happens when a plant wilts and dies from lack of water. This type of plasmolysis is more complicated compared to convex plasmolysis.

Examples of Plasmolysis

Plasmolysis is more common and happens in extreme cases of water loss. Some real-life examples of Plasmolysis are:

- Shrinkage of vegetables in hypertonic conditions.
- Blood cell shrinks when they are placed in the hypertonic conditions.
- During extreme coastal flooding, ocean water deposits salt onto land.

- Spraying of weedicides kills weeds in lawns, orchards and agricultural fields. This is due to the natural phenomena-Plasmolysis.
- When more amount of salt is added as the preservatives for food like jams, jellies, and pickles. The cells lose water due to higher concentration outside and become less conducive to support the growth of microorganisms.

Imbibition

Imbibition is a special kind of diffusion where water is diffused in by hydrophilic colloids, which otherwise do not dissolve in water. This results in an increase in the final volume. Imbibition in nature mostly relates to the movement of water. However, theoretically, any liquid can be imbibed. As per imbibition definition, the features of imbibition consists of two major players- the solid colloid substance called imbibants, and the liquid is called the imbibate.

For imbibition to occur, the imbibant needs to be hydrophilic. Hydrophobic substances cannot imbibe. For example, proteinaceous substances can imbibe while starchy substances do not. Therefore, this can be considered as the reason why proteinaceous pea seeds swell up in the water, but starchy wheat seeds do not.

Factors Affecting Imbibition

There are several factors affecting imbibition. Some of them are:

- Water Potential: Imbibants must have very high negative water potential. The more negative is the water potential of any substance, the greater is its capacity to imbibe. Water is considered to have the highest water potential, equals to zero (0) gradient.
- Water Potential Gradient: Water moves from a higher potential to a lower potential. The drier the imbibant is, the steeper is the water potential gradient. The water potential gradient determines the rate of imbibition.
- The Adsorption Capacity of the Imbibant: The adsorption capacity of the imbibant depends on its hydrophilicity. It is defined by the degree of attraction that the imbibant has for water molecules. It determines how much water can diffuse into the body and can be held by the imbibant molecules.
- The Affinity between Imbibant and Imbibate: The imbibant needs to have an affinity for the imbibate for imbibition to occur. If the imbibate molecules repel the molecule of the imbibant, imbibition will not occur.
- The pH of the System: The pH of the system plays a crucial role in the process of imbibition. It determines the affinity of the imbibant for the imbibate.
- The Temperature of the System: The temperature of the system determines the molecular spacing as well as the energy per molecule of any system. With an increase in temperature, the molecular spacing of the imbibant increases and thus increases the chances and extent of imbibition.
- The Pressure of the System: Pressure also determines the molecular spacing of any system. Therefore the pressure of the system determines the imbibition rate.

- The capacity of the imbibant to increase in volume: The imbibant must have the capacity to increase in volume after the imbibition process. If the imbibant has a rigid outer layer, it reduces the ability to increase in volume, and as a result, the rate of imbibition decreases.
- Heat of Wetting: Heat is produced during the process of imbibition. This heat is called the heat of wetting.

Conditions that are Necessary for Imbibition to Occur

There are three necessary conditions for imbibition to occur:

- A water potential gradient must be created between the imbibant and the imbibate. The imbibate moves from a region of higher water potential gradient to a region of lower water potential gradient.
- There should be some attraction and philicity between the imbibant and the imbibate.
- An increase in temperature increases the rate of imbibition.

Role of Imbibition in Plants

Imbibition has important roles to play in the process of plant development.

- Seed germination and initial stages of plant growth: Seed germination begins with the breaking of the seed coat. Water is imbibed when the seed coat is broken. After the seed coat is broken and the seedling grows, it needs to ascent through the soil above it. The growth of the seedling has a contribution from imbibition pressure developed within it.
- Absorption of water by roots: roots are the main tissues required for collecting water from the soil. Water moves into the root hair by imbibition. Imbibition also plays a major part in the ascent of sap through the plant body.
- The plant keeps its cells moist by transferring water from adjacent cells through imbibition.

Imbibition Pressure

When the imbibant is kept in a conned area, pressure increases with an increase in the imbibant volume. This pressure is known as imbibition pressure. It is also known as the matric potential since it develops due to the matric potential of the imbibing substance. It is denoted as Ψm and is measured in megapascals (MPa) or bars.

This imbibition pressure can be tremendous and can be used to split open a rock. For example, in ancient times, Egyptians used to put a dry stick in any crevices of the rock and used to apply water on the stick. The rock used to split open due to the constant imbibition pressure from the water.

Significance of Imbibition in Nature

- Imbibition helps in seed germination.
- Imbibition helps in water absorption by root hairs of plants.
- Imbibition helps to keep the cells moist.

Although imbibition is a form of diffusion, there are certain subtle differences between the two.

Table: Difference between diffusion and Imbibition.

Diffusion	Imbibition
Diffusion involves movement of any substance from a region of its higher concentration to a region of its lower concentration.	Imbibition is the movement of a solvent (mainly water) from a region of its higher concentration to a region of its lower concentration.
No adsorption is involved.	Adsorption is involved.
No change in volume.	The volume of the imbibant increases.
No affinity is required.	Affinity must be present between the imbibant and the imbibate.
Little change in pressure.	An enormous increase in pressure.
No heat is liberated.	Heat is liberated.
The Diffusion of gases can also take place.	Gas is not involved.

Water Potential

Water potential is the energy required, per quantity of water, to transport an infinitesimal quantity of water from the sample to a reference pool of pure free water. To understand what that means, compare the water in a soil sample to water in a drinking glass. The water in the glass is relatively free and available; the water in the soil is bound to surfaces diluted by solutes and under pressure or tension. In fact, the soil water has a different energy state from "free" water. The free water can be accessed without exerting any energy. The soil water can only be extracted by expending energy. Soil water potential expresses how much energy you would need to expend to pull that water out of the soil sample.

Soil water potential is a differential property. For the measurement to have meaning, a reference must be specified. The reference typically specified is pure, free water at the soil surface. The water potential of this reference is zero. Water potential in the environment is almost always less than zero, because you have to add energy to get the water out.

Intensive vs. Extensive Variables

Water movement in the environment is really a physics problem, and to understand it, we have to distinguish between intensive and extensive variables. The extensive variable describes the extent or amount of matter or energy. The intensive variable describes the intensity or quality of matter or energy. For example, the thermal state of a substance can be described in terms of both heat content and temperature.

The two variables are related, but they are not the same. Heat content depends on mass, specific heat, and temperature. You can't tell by measuring heat content whether or not heat will be transferred to another object if the two touch each other. So you also don't know if the object is hot or cold or whether it will be safe to touch.

These questions are much easier to answer if you know the intensive variable—temperature. In fact, though it can be important to measure both intensive and extensive variables, often the intensive variable gives you more useful information. In terms of water, the extensive variable is water content, and it tells you the extent, or amount, of water in plant tissue or soil. The intensive variable is water potential, and it describes the intensity or quality of water in plant tissue or soil. Many questions about water availability and movement are best answered by measuring soil water potential.

Water Movement

Water will always flow from high potential to low potential. This is the second law of thermodynamics—energy flows along the gradient of the intensive variable. Water will move from a higher energy location to a lower energy location until the locations reach equilibrium, as illustrated in Figure 1. For example, if a soil's water potential were -50 kPa, water would move toward the more negative -100 kPa to become more stable.

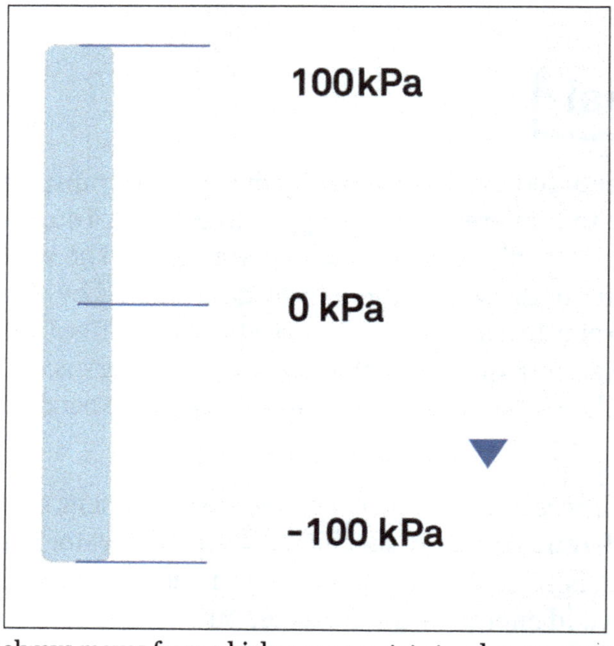

Water always moves from a higher energy state to a lower energy state.

Plant Water Availability

Liquid water moves from soil to and through roots, through the xylem of plants, to the leaves, and eventually evaporates in the sub-stomatal cavities of the leaf. The driving force for this flow is a water potential gradient. Thus, in order for water to flow, the leaf water potential must be lower than the soil water potential. In Figure below, the soil is at -0.3 MPa and the roots are slightly more negative at -0.5 MPa. This means the roots will pull water up from the soil. Then the water will move up through the xylem and out through the leaves. And the atmosphere, at -100 MPa, is what drives this gradient.

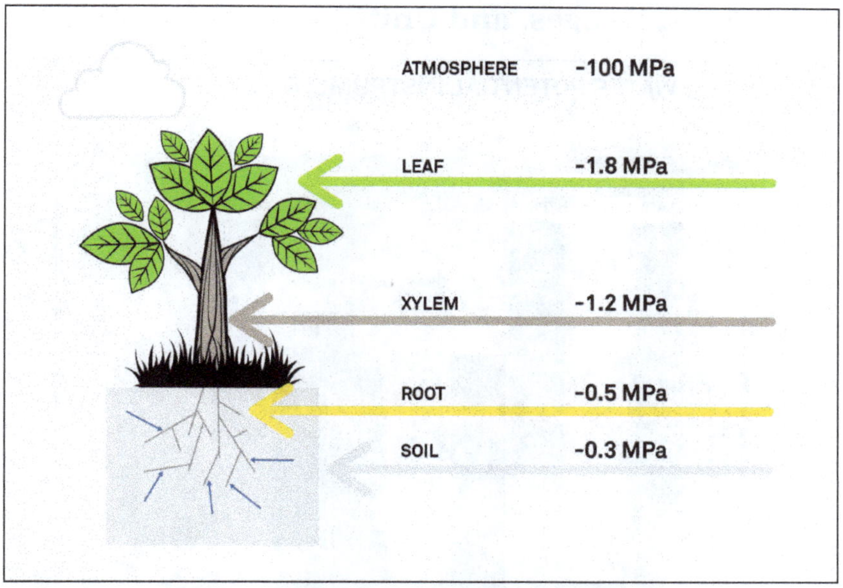

Example water potential gradient in a system.

Water potential measurements clearly indicate plant available water, and unlike water content, there is an easy reference scale–plant optimal runs from about -2-5 kPa which is on the very wet side, to approximately -100 kPa, at the drier end of optimal. Below that, plants will be in deficit, and past -1000 kPa they start to suffer. Depending on the plant, water potentials below -1000 to -2000 kPa cause permanent wilting.

Irrigators and scientists use water potential sensors in conjunction with water content sensors to understand plant water availability. In Figure 3, you can observe where the water content declines and at what percentage the plants begin to stress. It's also possible to recognize when the soil has too much water: the water content is above where water potential sensors start to sense plant stress. Using this information, researchers can identify the plant optimal range at 12% to 17% volumetric water content. Anything below or above that range will be too little or too much water.

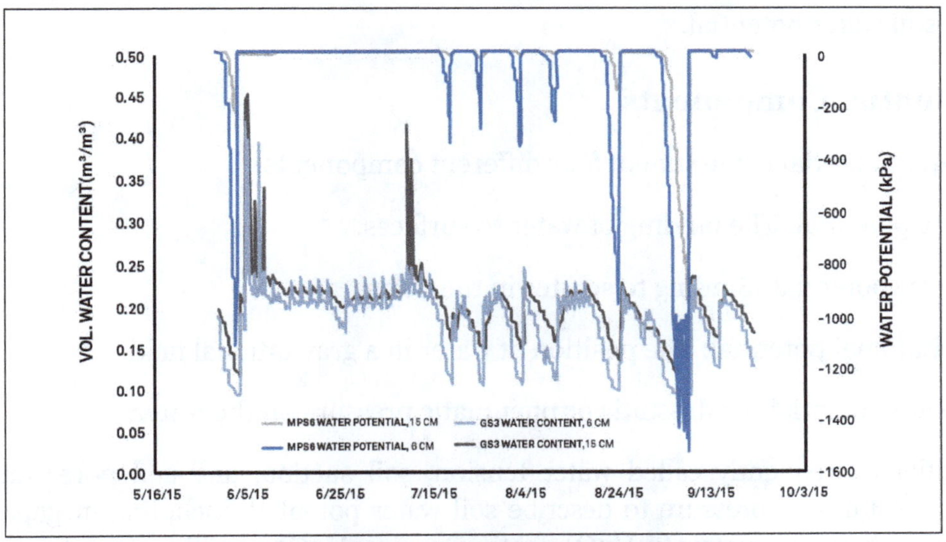

Water content and water potential data for turf grass.

Water Potential Names, Ranges, and Units

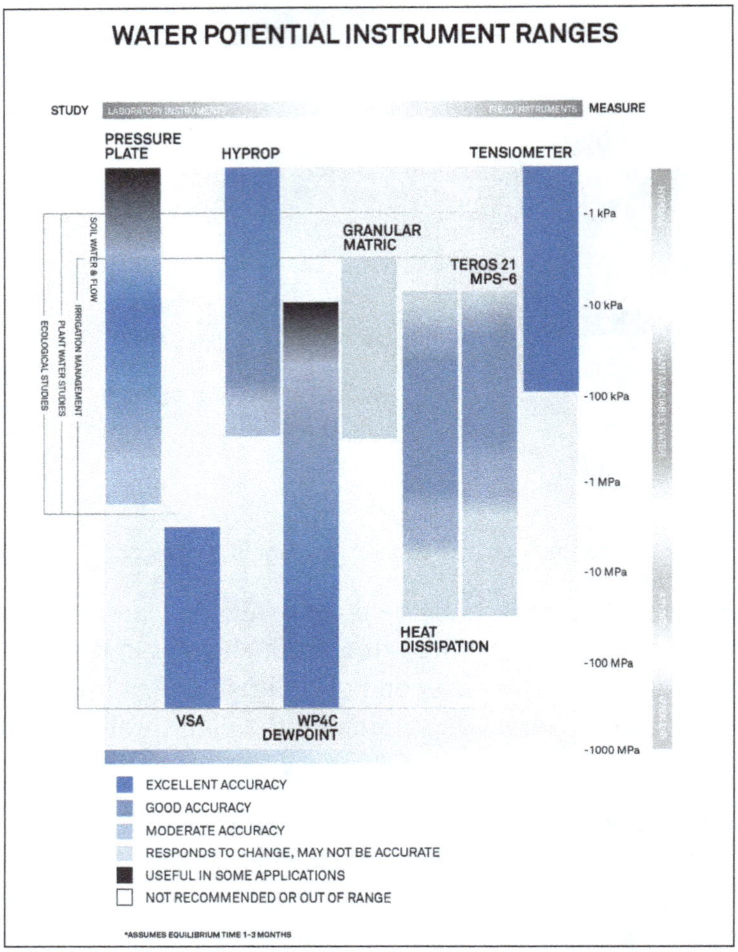

A range comparison of water potential instruments.

Figure above illustrates that there are different water potential instruments that measure different ranges. Watch the video to see how you can combine METER LABROS instruments to measure the full range of soil water potential.

Water Potential Components

The total water potential is the sum of four different components:

- Matric potential: The binding of water to surfaces.
- Osmotic potential: Binding to solutes in the water.
- Gravitational potential: The position of water in a gravitational field.
- Pressure potential: Hydrostatic or pneumatic pressure on the water.

Water potential is frequently called water tension, soil suction, and soil pore water pressure. We typically use units of pressure to describe soil water potential, including megapascals (MPa), kilopascals (kPa), bars, and meters (mH2O), centimeters (cm H2O), or millimeters of water (mmH2O).

Water potential is actually measured in energy per unit of mass, so the official units should be joules per kilogram, but if you take into account the density of water, the units become kilopascals; therefore we typically describe it using units of pressure.

How to Calculate Water Potential

Soil water potential is the sum of four different components: gravitational potential + the matric potential + the pressure potential + the osmotic potential.

$$\Psi = \Psi_g + \Psi_m + \Psi_p + \Psi_o$$

Matric potential is the most significant component as far as soil is concerned because it relates to the water that is adhering to soil surfaces. In Figure 5, the matric potential is what created the water film clinging to the soil particles. As water drains out of the soil, the air-filled pore spaces get bigger, and the water gets more tightly bound to the soil particles as the matric potential decreases.

Matric Potential

Matric potential arises because water is attracted to most surfaces through hydrogen bonding and van der Waals forces. Soil is made up of small particles, providing lots of surfaces that will bind water. This binding is highly dependent on soil type. For example, sandy soil has large particles which provide less surface-binding sites, while a silt loam has smaller particles and more surface-binding sites.

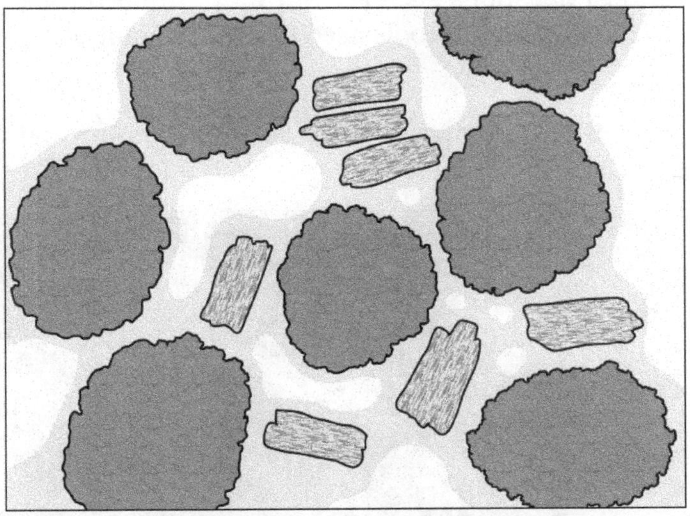

Cross-sectional diagram of a soil. As the soil adsorbs water, it creates a water film that clings to the soil particles. The matric potential is what creates the water film.

The following figure, showing moisture release curves for three different types of soil, demonstrates the effect of surface area. Sand, containing 10% water, has a high matric potential, and the water is readily available to organisms and plants. Silt loam, containing 10% water, will have a much lower matric potential, and the water will be significantly less available.

Matric potential is always negative or zero and is the most significant component of soil water potential in unsaturated conditions.

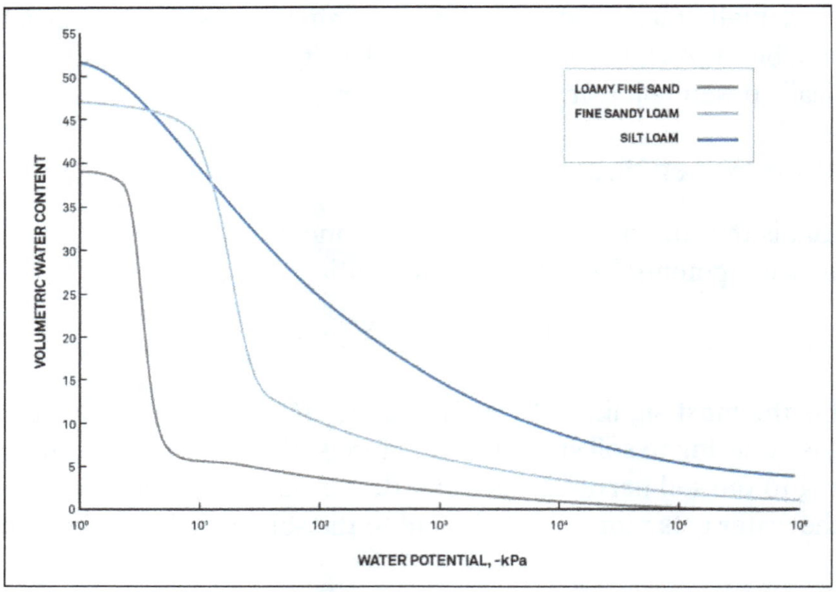

Moisture release curves for three different types of soil demonstrate the effect of surface area.

A Soil Moisture Power Tool

Soil moisture release curves (soil-water characteristic curves) illustrate the relationship between water potential and water content and are like physical fingerprints, unique to each soil type. Use them in your research to understand and predict the fate of water in your particular soil. Moisture release curves answer critical questions such as: will water drain through the soil quickly or be held in the root zone? They are powerful tools used to predict plant water uptake, deep drainage, runoff, and more.

Tensiometers and the TEROS 21 are both soil water potential sensors that measure matric potential in the field.

TEROS 21 field sensor measures matric potential which indicates water availability.

Osmotic Potential

Osmotic potential describes the dilution and binding of water by solutes that are dissolved in the water. This potential is also always negative. Osmotic potential only affects the system if there is a semi-permeable barrier that blocks the passage of solutes. This is actually quite common in nature. For example, plant roots allow water to pass but block most solutes. Cell membranes also form a semi-permeable barrier. A less-obvious example is the air-water interface, where water can pass into air in the vapor phase but salts are left behind.

You can calculate osmotic potential from the following equation if you know the concentration of solute in the water:

$$\Psi_o = C\Phi VRT$$

Where *C* is the concentration of solute (mol/kg), ϕ is the osmotic coefficient (-0.9 to 1 for most solutes), *v* is the number of ions per mol (NaCl= 2, $CaCl_2$ = 3, sucrose = 1), *R* is the gas constant, and *T* is the Kelvin temperature.

Osmotic potential is always negative or zero and is significant in plants and some salt-affected soils.

Gravitational Potential

Gravitational potential arises because of water's location in a gravitational field. It can be positive or negative, depending on where you are in relation to the specified reference of pure, free water at the soil surface. Gravitational potential is then:

$$\Psi_G = GH$$

Where *G* is the gravitational constant (9.8 m s^{-2}) and *H* is the vertical distance from the reference height to the soil surface (the specified height).

Pressure Potential

Pressure potential is a hydrostatic or pneumatic pressure being applied to or pulled on the water. It is more macroscopic effect acting throughout a larger region of the system.

There are several examples of positive pressure potential in the natural environment. For example, there is a positive pressure present below the surface of any groundwater. You can feel this pressure yourself as you swim down into a lake or pool. Similarly, a pressure head or positive pressure potential develops as you move below the water table. Turgor pressure in plants and blood pressure in animals are two more examples of positive pressure potential.

Pressure potential can be calculated from:

$$\Psi = P / P_w$$

Where P is the pressure (P_a) and P_W is the density of water.

Though pressure potential is usually positive, there are important cases where it is not. One is found in plants, where a negative pressure potential in the xylem draws water from the soil up through the roots and into the leaves.

Water Potential and Relative Humidity

Water potential and relative humidity are related by the Kelvin equation. If you know temperature and humidity, you can calculate the water potential using this equation:

$$\Psi = RT / M_w "LN( )"H_R$$

Where Ψ is water potential (MPa), H_R is the relative humidity (unit less), R is the universal gas constant (8.3143 J mol^{-1} K^{-1}), M_W is the mass of water (18.02 g/mol), and T is Kelvin temperature.

Water Transportation

Water is the most limiting abiotic (non-living) factor to plant growth and productivity, and a principal determinant of vegetation distributions worldwide. Since antiquity, humans have recognized plants' thirst for water as evidenced by the existence of irrigation systems at the beginning of recorded history. Water's importance to plants stems from its central role in growth and photosynthesis, and the distribution of organic and inorganic molecules. Despite this dependence, plants retain less than 5% of the water absorbed by roots for cell expansion and plant growth. The remainder passes through plants directly into the atmosphere, a process referred to as transpiration. The amount of water lost via transpiration can be incredibly high; a single irrigated corn plant growing in Kansas can use 200 L of water during a typical summer, while some large rainforest trees can use nearly 1200 L of water in a single day.

If water is so important to plant growth and survival, then why would plants waste so much of it? The answer to this question lies in another process vital to plants — photosynthesis. To make sugars, plants must absorb carbon dioxide (CO_2) from the atmosphere through small pores in their leaves called stomata. However, when stomata open, water is lost to the atmosphere at a prolific rate relative to the small amount of CO_2 absorbed; across plant species an average of 400 water molecules are lost for each CO_2 molecule gained. The balance between transpiration and photosynthesis forms an essential compromise in the existence of plants; stomata must remain open to build sugars but risk dehydration in the process.

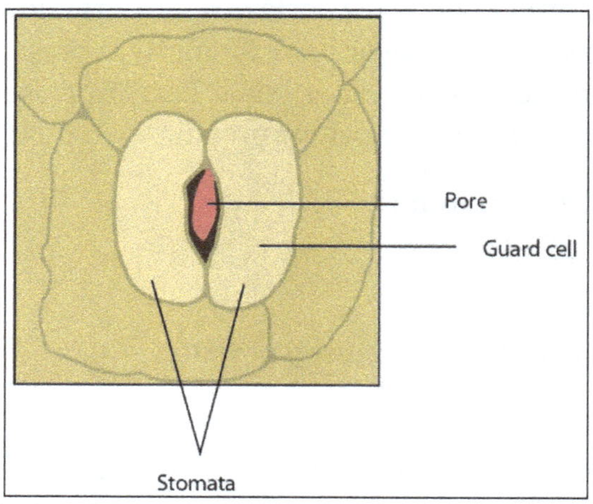

Rendering of an open stoma on the surface of a tobacco leaf.

Stomata are pores found on the leaf surface that regulate the exchange of gases between the leaf's interior and the atmosphere. Stomatal closure is a natural response to darkness or drought as a means of conserving water.

From the Soil into the Plant

Essentially all of the water used by land plants is absorbed from the soil by roots. A root system consists of a complex network of individual roots that vary in age along their length. Roots grow from their tips and initially produce thin and non-woody fine roots. Fine roots are the most permeable portion of a root system, and are thought to have the greatest ability to absorb water, particularly in herbaceous (i.e., non-woody) plants. Fine roots can be covered by root hairs that significantly increase the absorptive surface area and improve contact between roots and the soil. Some plants also improve water uptake by establishing symbiotic relationships with mycorrhizal fungi, which functionally increase the total absorptive surface area of the root system.

Root hairs often form on fine roots and improve water absorption by increasing root surface area and by improving contact with the soil.

Roots of woody plants form bark as they age, much like the trunks of large trees. While bark formation decreases the permeability of older roots they can still absorb considerable amounts of water. This is important for trees and shrubs since woody roots can constitute ~99% of the root surface in some forests.

Roots have the amazing ability to grow away from dry sites toward wetter patches in the soil — a phenomenon called hydrotropism. Positive hydrotropism occurs when cell elongation is inhibited on the humid side of a root, while elongation on the dry side is unaffected or slightly stimulated resulting in a curvature of the root and growth toward a moist patch. The root cap is most likely the site of hydro sensing; while the exact mechanism of hydrotropism is not known, recent work with the plant model *Arabidopsis* has shed some light on the mechanism at the molecular level.

Roots of many woody species have the ability to grow extensively to explore large volumes of soil. Deep roots (>5 m) are found in most environments allowing plants to access water from permanent water sources at substantial depth. Roots from the Shepard's tree (*Boscia albitrunca*) have been found growing at depths 68 m in the central Kalahari, while those of other woody species can spread laterally up to 50 m on one side of the plant. Surprisingly, most arid-land plants have very shallow root systems, and the deepest roots consistently occur in climates with strong seasonal precipitation (i.e., Mediterranean and monsoonal climates).

Tree roots at significant depths accessed via caves.

Plant scientists examine: deep roots of Juniperus asheii growing at 7m depth in a cave in Austin, TX USA (left); an extensive fine root network attached to a single ~1cm diameter tap root accessing a perennial underground stream at 20m depth in a cave in central TX, USA; and twisty roots in a cave located in southwest Western Australia below a forest dominated by Eucalyptus diversicolor — roots in this cave system are commonly found from 20-60m depth.

Through the Plant into the Atmosphere

Water flows more efficiently through some parts of the plant than others. For example, water absorbed by roots must cross several cell layers before entering the specialized water transport tissue (referred to as xylem). These cell layers act as a filtration system in the root and have a much greater resistance to water flow than the xylem, where transport occurs in open tubes. Imagine the difference between pushing water through numerous coffee filters versus a garden hose. The relative ease with which water moves through a part of the plant is expressed quantitatively using the following equation:

$$\text{Flow} = \Delta\psi / R,$$

Which is analogous to electron flow in an electrical circuit described by Ohm's law equation:

$$i = V / R,$$

Where R is the resistance, i is the current or flow of electrons, and V is the voltage. In the plant system, V is equivalent to the water potential difference driving flow ($\Delta\psi$) and i is equivalent to the flow of water through/across a plant segment. Using these plant equivalents, the Ohm's law analogy can be used to quantify the hydraulic conductance (i.e., the inverse of hydraulic R) of individual segments (i.e., roots, stems, leaves) or the whole plant (from soil to atmosphere).

Upon absorption by the root, water first crosses the epidermis and then makes its way toward the center of the root crossing the cortex and endodermis before arriving at the xylem. Along the way, water travels in cell walls (apoplastic pathway) and/or through the inside of cells (cell to cell pathway, C-C). At the endodermis, the apoplastic pathway is blocked by a gasket-like band of suberin — a waterproof substance that seals off the route of water in the apoplast forcing water to cross via the C-C pathway. Because water must cross cell membranes (e.g., in the cortex and at apoplastic barriers), transport efficiency of the C-C pathway is affected by the activity, density, and location of water-specific protein channels embedded in cell membranes (i.e., aquaporins). Much work over the last two decades has demonstrated how aquaporins alter root hydraulic resistance and responds to abiotic stress, but their exact role in bulk water transport is yet unresolved.

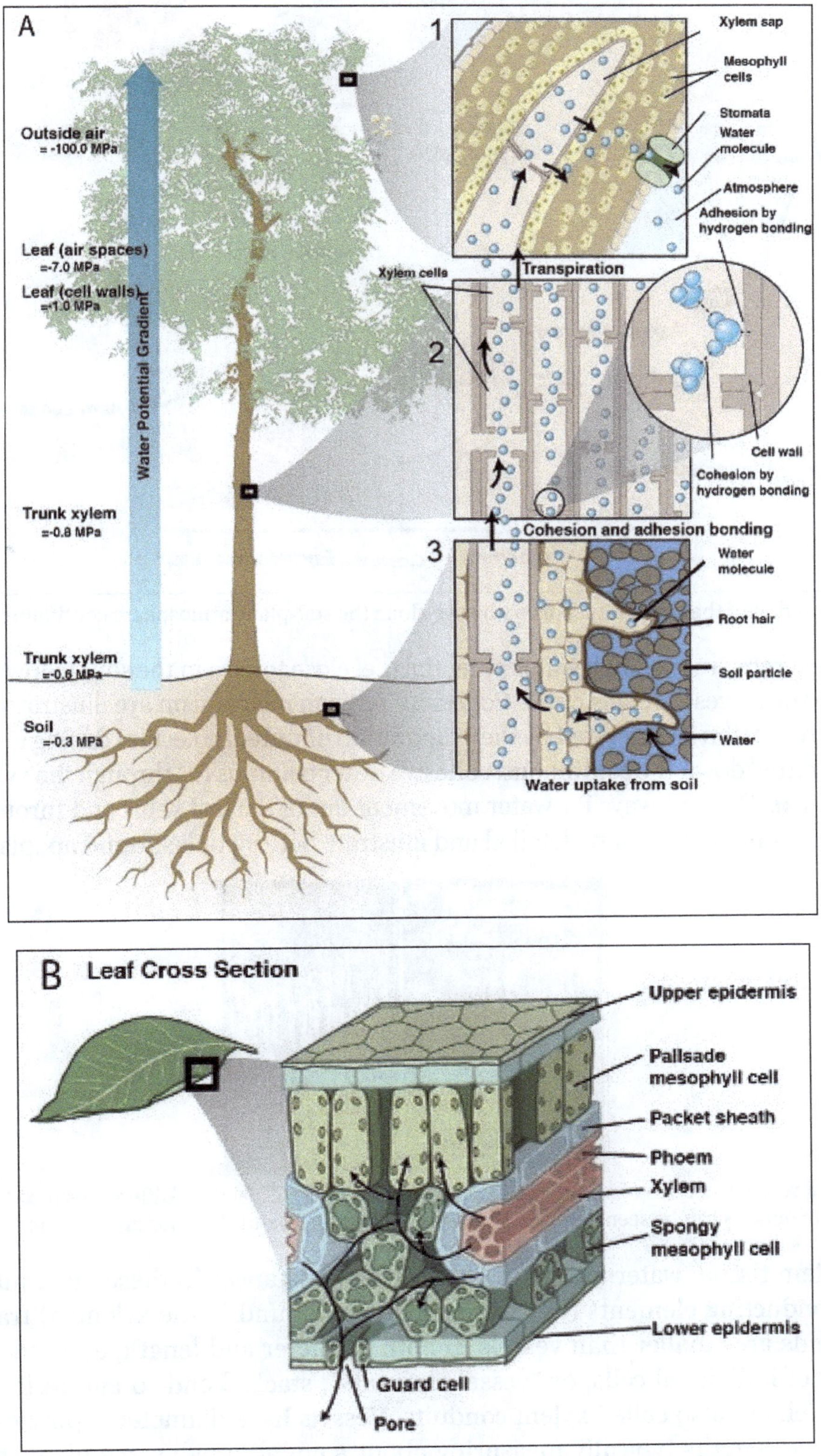

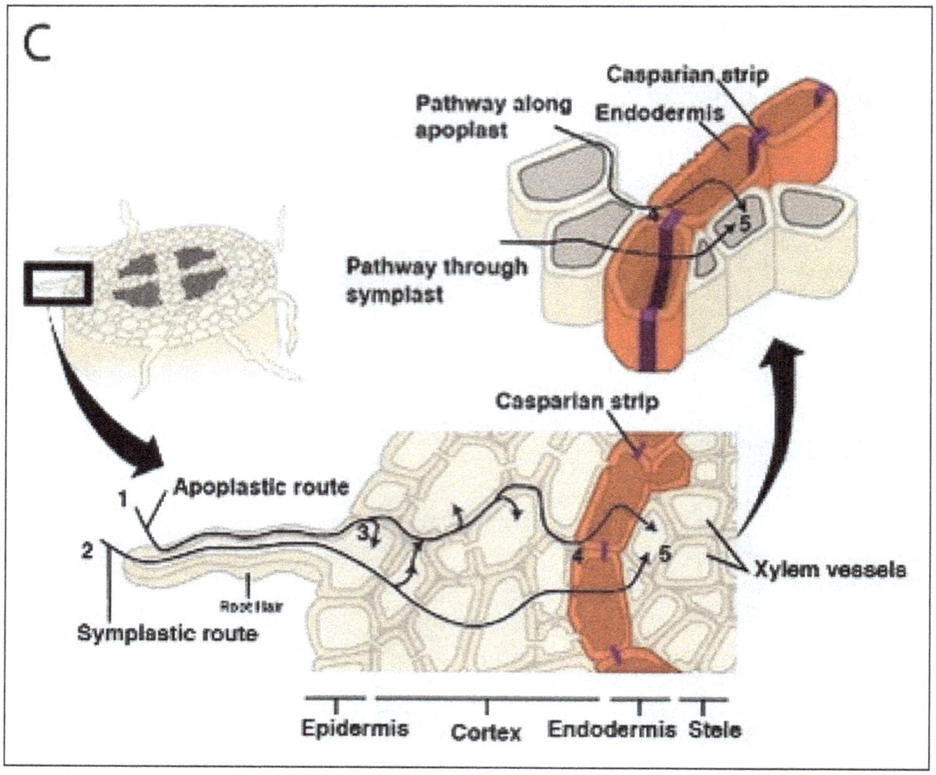

Representation of the water transport pathways along the soil-plant-atmosphere continuum (SPAC).

(A) Water moves from areas of high water potential (i.e. close to zero in the soil) to low water potential (i.e., air outside the leaves). Details of the Cohesion-Tension mechanism are illustrated with the inset panels (A), where tension is generated by the evaporation of water molecules during leaf transpiration (1) and is transmitted down the continuous, cohesive water columns (2) through the xylem and out the roots to the soil (3). The pathways for water movement out of the leaf veins and through the stomata (B) and across the fine roots (C) are detailed and illustrate both symplastic and apoplastic pathways.

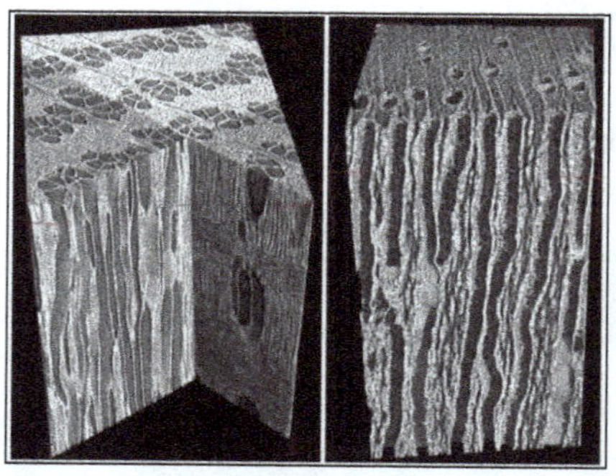

Three dimensional reconstructions of xylem imaged at the Ghent microCT facility. Differences in xylem structure and conduit distributions can be seen between Ulmus americana (left) and Fraxinus americana (right) xylem.

Once in the xylem tissue, water moves easily over long distances in these open tubes. There are two kinds of conducting elements (i.e., transport tubes) found in the xylem: 1) tracheids and 2) vessels. Tracheids are smaller than vessels in both diameter and length, and taper at each end. Vessels consist of individual cells, or "vessel elements", stacked end-to-end to form continuous open tubes, which are also called xylem conduits. Vessels have diameters approximately that of a human hair and lengths typically measuring about 5 cm although some plant species contain vessels as long as 10 m. Xylem conduits begin as a series of living cells but as they mature the cells commit suicide (referred to as programmed cell death), undergoing an ordered deconstruction where they lose their cellular contents and form hollow tubes. Along with the water conducting tubes, xylem tissue contains fibers which provide structural support, and living metabolically-active parenchyma cells that are important for storage of carbohydrates, maintenance of flow within a conduit (see details about embolism repair below), and radial transport of water and solutes.

When water reaches the end of a conduit or passes laterally to an adjacent one, it must cross through pits in the conduit cell walls (Figure 6). Bordered pits are cavities in the thick secondary cell walls of both vessels and tracheids that are essential components in the water-transport system of higher plants. The pit membrane, consisting of a modified primary cell wall and middle lamella, lies at the center of each pit, and allows water to pass between xylem conduits while limiting the spread of air bubbles (i.e., embolism) and xylem-dwelling pathogens. Thus, pit membranes function as safety valves in the plant water transport system. Averaged across a wide range of species, pits account for >50% of total xylem hydraulic resistance. The structure of pits varies dramatically across species, with large differences evident in the amount of conduit wall area covered by pits, and in the porosity and thickness of pit membranes.

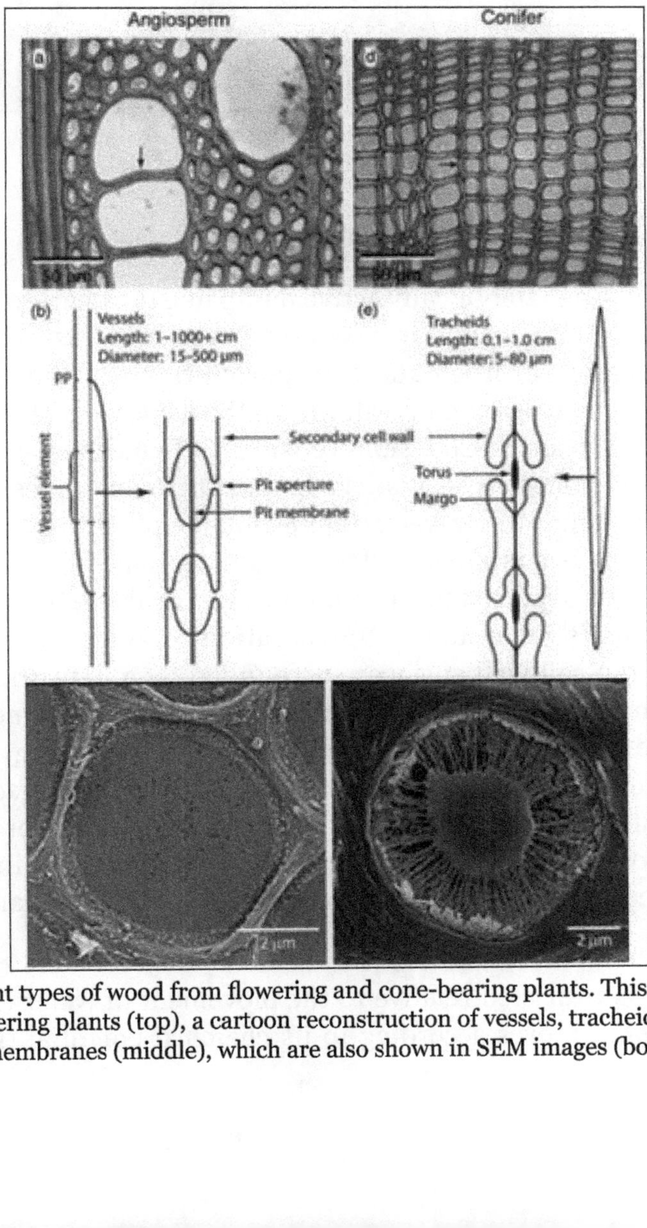

Comparison of different types of wood from flowering and cone-bearing plants. This features wider conduits from flowering plants (top), a cartoon reconstruction of vessels, tracheids and their pit membranes (middle), which are also shown in SEM images (bottom).

After traveling from the roots to stems through the xylem, water enters leaves via petiole (i.e., the leaf stalk) xylem that branches off from that in the stem. Petiole xylem leads into the mid-rib (the main thick vein in leaves), which then branch into progressively smaller veins that contain tracheids and are embedded in the leaf mesophyll. In dicots, minor veins account for the vast majority of total vein length, and the bulk of transpired water is drawn out of minor veins. Vein arrangement, density, and redundancy are important for distributing water evenly across a leaf, and may buffer the delivery system against damage (i.e., disease lesions, herb ivory, air bubble spread). Once water leaves the xylem, it moves across the bundle sheath cells surrounding the veins. It is still unclear the exact path water follows once it passes out of the xylem through the bundle sheath cells and into the mesophyll cells, but is likely dominated by the apoplastic pathway during transpiration.

An example of a venation pattern to illustrate the hydraulic pathway from petiole xylem into the leaf cells and out the stomata.

Mechanism Driving Water Movement in Plants

Unlike animals, plants lack a metabolically active pump like the heart to move fluid in their vascular system. Instead, water movement is passively driven by pressure and chemical potential gradients. The bulk of water absorbed and transported through plants is moved by negative pressure generated by the evaporation of water from the leaves (i.e., transpiration) — this process is commonly referred to as the Cohesion-Tension (C-T) mechanism. This system is able to function because water is "cohesive" — it sticks to itself through forces generated by hydrogen bonding. These hydrogen bonds allow water columns in the plant to sustain substantial tension (up to 30 MPa when water is contained in the minute capillaries found in plants), and helps explain how water can be transported to tree canopies 100 m above the soil surface. The tension part of the C-T mechanism is generated by transpiration. Evaporation inside the leaves occurs predominantly from damp cell wall surfaces surrounded by a network of air spaces. Menisci form at this air-water interface, where apoplastic water contained in the cell wall capillaries is exposed to the air of the sub-stomatal cavity. Driven by the sun's energy to break the hydrogen bonds between molecules, water evaporates from menisci, and the surface tension at this interface pulls water molecules to replace those lost to evaporation. This force is transmitted along the continuous water columns down to the roots, where it causes an influx of water from the soil. Scientists call the continuous water transport pathway the Soil Plant Atmosphere Continuum (SPAC).

When solute movement is restricted relative to the movement of water (i.e., across semipermeable cell membranes) water moves according to its chemical potential (i.e., the energy state of water) by osmosis — the diffusion of water. Osmosis plays a central role in the movement of water between cells and various compartments within plants. In the absence of transpiration, osmotic forces dominate the movement of water into roots. This manifests as root pressure and guttation — a process commonly seen in lawn grass, where water droplets form at leaf margins in the morning after conditions of low evaporation. Root pressure results when solutes accumulate to a greater concentration in root xylem than other root tissues. The resultant chemical potential gradient drives water influx across the root and into the xylem. No root pressure exists in rapidly transpiring plants, but it has been suggested that in some species root pressure can play a central role in the refilling of non-functional xylem conduits particularly after winter.

Disruption of Water Movement

Water transport can be disrupted at many points along the SPAC resulting from both biotic and abiotic factors. Root pathogens (both bacteria and fungi) can destroy the absorptive surface area in the soil, and similarly foliar pathogens can eliminate evaporative leaf surfaces, alter stomatal function, or disrupt the integrity of the cuticle. Other organisms (i.e., insects and nema-todes) can cause similar disruption of above and below ground plant parts involved in water trans-port. Biotic factors responsible for ceasing flow in xylem conduits include: pathogenic organisms and their by-products that plug conduits; plant-derived gels and gums produced in response to pathogen invasion; and tyloses, which are outgrowths produced by living plant cells surrounding a vessel to seal it off after wounding or pathogen invasion.

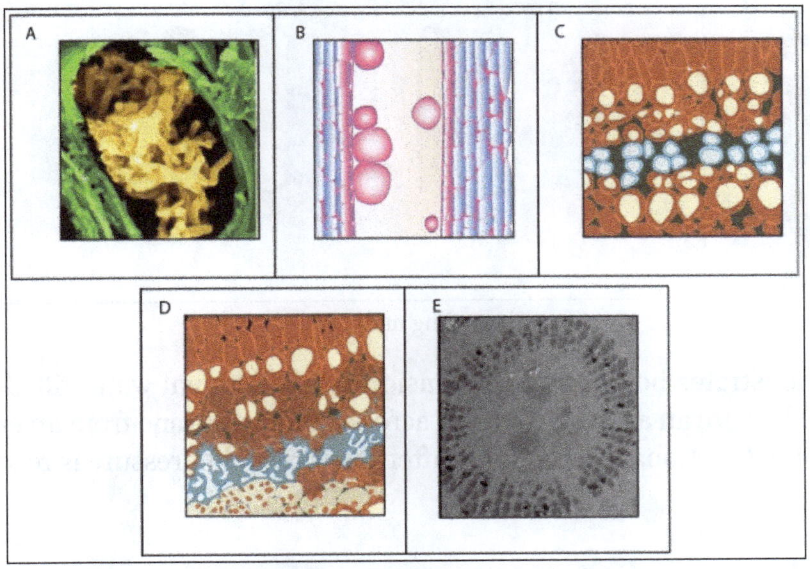

Sources of dysfunction in the xylem.

Left to right: (A) xylem-dwelling pathogens like Xylella fastidiosa bacteria; (B) tyloses (plant-derived); (C and D) conduit (in blue) implosion; and (E) embolized conduits among water filled ones in a frozen plant samples.

Abiotic factors can be equally disruptive to flow at various points along the water transport pathway. During drought, roots shrink and lose contact with water adhering to soil particles — a process that can also be beneficial by limiting water loss by roots to drying soils (i.e., water can flow in reverse and leak out of roots being pulled by drying soil). Under severe plant dehydration, some pine needle conduits can actually collapse as the xylem tensions increase.

Water moving through plants is considered meta-stable because at a certain point the water column breaks when tension becomes excessive — a phenomenon referred to as cavitation. After cavitation occurs, a gas bubble (i.e., embolism) can form and fill the conduit, effectively blocking water movement. Both sub-zero temperatures and drought can cause embolisms. Freezing can induce embolism because air is forced out of solution when liquid water turns to ice. Drought also induces embolism because as plants become drier tension in the water column increases. There is a critical point where the tension exceeds the pressure required to pull air from an empty conduit to a filled conduit across a pit membrane — this aspiration is known as air seeding. An air seed creates a void in the water, and the tension causes the void to expand and break the continuous column. Air seeding thresholds are set by the maximum pore diameter found in the pit membranes of a given conduit.

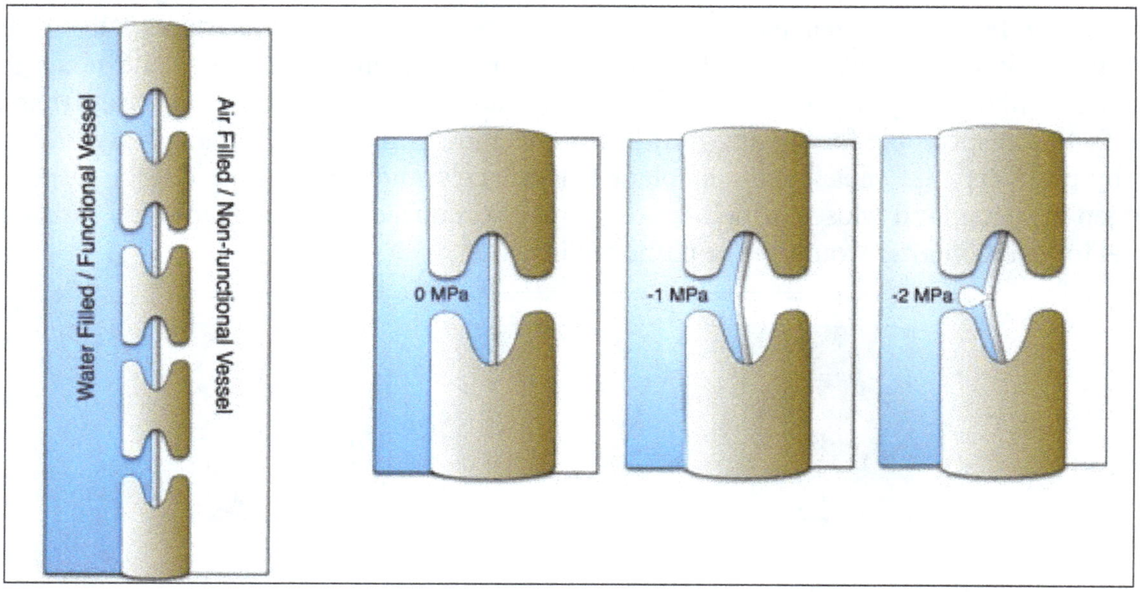

Air seeding mechanism.

Figure above demonstrates how increasing tension in a functional water filled vessel eventually reaches a threshold where an air seed is pulled across a pit membrane from an embolized conduit. Air is seeded into the functional conduit only after the threshold pressure is reached.

Water Absorption

In higher plants water is absorbed through root hairs which are in contact with soil water and form a root hair zone a little behind the root tips. Root hairs are tubular hair like prolongations of the cells of the epidermal layer (when epidermis bears root hairs it is also known as piliferous layer) of the roots. The walls of root hairs are permeable and consist of pectic substances and cellulose which are strongly hydrophilic (water loving) in nature. Root hairs contain vacuoles filled with cell sap.

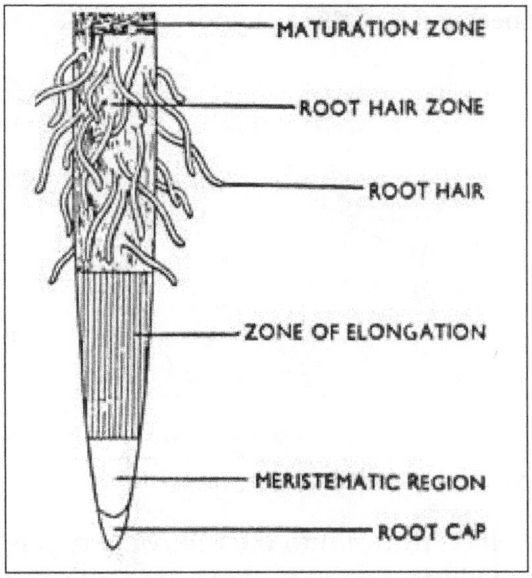

Diagrammatic representation of a root tip showing root hair zone.

When roots elongate, the older hairs die and new root hairs are developed so that they are in contact with fresh supplies of water in the soil.

Mechanism of Water Absorption is of Two Types

Active Absorption of Water

In this process the root cells play active role in the absorption of water and metabolic energy released through respiration is consumed.

Active absorption may be of two kinds:

- Osmotic absorption i.e., when water is absorbed from the soil into the xylem of the roots according to the osmotic gradient.
- Non-osmotic absorption i.e., when water is absorbed against the osmotic gradient.

Active Osmotic Absorption of Water

First step in the osmotic absorption of water is the imbibition of soil water by the hydrophilic cell walls of root hairs. Osmotic Pressure (O.P.) of the cell-sap of root hairs is usually higher than the O.P. of the soil water. Therefore, the Diffusion Pressure Deficit (D.P.D.) and the suction pressure in the root hairs become higher and water from the cell walls enters into them through plasma-membrane (semi-permeable) by osmotic diffusion. As a result, the O.P., suction pressure and D.P.D. of root hairs now become lower, while their turgor pressure is increased.

Now, the cortical cells adjacent to root hairs have higher O.P., suction pressure and D.P.D. in comparison to the root hairs. Therefore, water is drawn into the adjacent cortical cells from the root-hairs by osmotic diffusion.

In the same way, the water by cell to cell osmotic diffusion gradually reaches the innermost cortical cells and the endodermis. Osmotic diffusion of water into endodermis takes place through special thin walled passage cells because the other endodermal cells have casparian strips on their walls which are impervious to water.

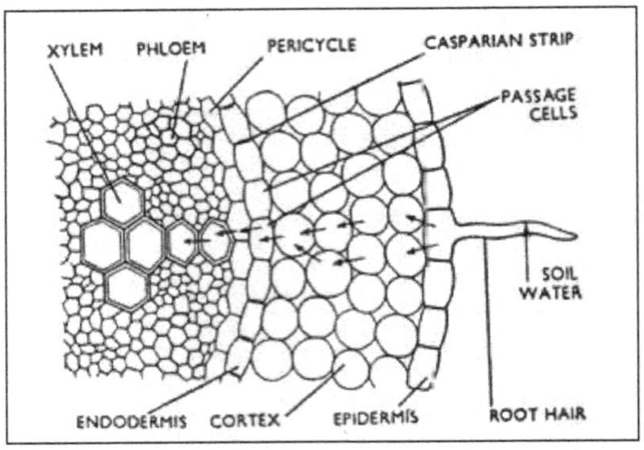

A part of T.S. of typical dicot root. The arrows indicate the path of water.

Water from endodermal cells is drawn into the cells of pericycle by osmotic diffusion which now becomes turgid and their suction pressure is decreased. In the last step, water is drawn into xylem from turgid pericycle cells. (In roots the vascular bundles are radial and protoxylem elements are in contact with pericycle).

It is because in absence of turgor pressure of the xylem vessels (which are non-elastic), the suction pressure of xylem vessels become higher than the suction pressure of the cells of the pericycle. When water enters into xylem from pericycle, a pressure is developed in the xylem of roots which can raise the water to a certain height in the xylem. This pressure is called as root pressure.

Active Non-Osmotic Absorption of Water

Sometimes, it has been observed that absorption of water takes place even when the O.P. of the soil water is higher than the O.P. of cell-sap. This type of absorption which is non- osmotic and against the osmotic gradient requires the expenditure of metabolic energy probably through respiration.

Following evidences support this view:

- The factors which inhibit respiration also decrease water absorption.
- Poisons which retard metabolic activities of the root cells also retard water absorption.
- Auxins (growth hormones) which increase metabolic activities of the cells stimulate absorption of water.

Passive Absorption of Water

Passive absorption of water takes place when rate of transpiration is usually high. Rapid evaporation of water from the leaves during transpiration creates a tension in water in the xylem of the leaves. This tension is transmitted to water in xylem of roots through the xylem of stem and the water rises upward to reach the transpiring surfaces.

As a result, soil water enters into the cortical cells through root hairs to reach the xylem of roots to maintain the supply of water. The force for this entry of water is created in leaves due to rapid transpiration and hence, the root cells remain passive during this process.

During absorption of water by roots, the flow of water from epidermis to endodermis may take place through three different pathways:

- Apoplastic pathway (cell walls and intercellular spaces).
- Trans-membrane pathway (by crossing the plasma membranes).
- Symplast pathway (through plasmodesmata).

The mechanism of water absorption described earlier, in-fact belongs to the second category. The relative importance of these three pathways in water absorption by roots is not clearly established. However, a combination of these three pathways is responsible for transport of water across the root.

External Factors Affecting Absorption of Water

Available Soil Water

Sufficient amount of water should be present in the soil in such form which can easily be absorbed by the plants. Usually the plants absorb capillary water i.e., water present in films in between soil particles. Other forms of water in the soil e.g., hygroscopic water, combined-water, gravitational water etc. are not easily available to plants. Increased amount of water in the soil beyond a certain limit results in poor aeration of the soil which retards metabolic activities of root cells like respiration and hence, the rate of water absorption is also retarded.

Concentration of the Soil Solution

Increased conc. of soil solution (due to the presence of more salts in the soil) results in higher osmotic pressure. If the O.P. of soil solution will become higher than the O.P. of cell sap in root cells, the water absorption particularly the osmotic absorption of water will be greatly suppressed. Therefore, absorption of water is poor in alkaline soils and marshes.

Soil Air

Absorption of water is retarded in poorly aerated soils because in such soils deficiency of O1 and consequently the accumulation of CO_2 will retard the metabolic activities of the roots like respiration. This also inhibits rapid growth and elongation of the roots so that they are deprived of the fresh supply of water in the soil. Water logged soils are poorly aerated and hence, are physiologically dry. They are not good for absorption of water.

Soil Temperature

Increase in soil temperature up to about 30°C favors water absorption. At higher temperatures water absorption is decreased. At low temp, also water absorption decreases so much so that at about 0°C it is almost checked.

This is probably because at low temp:

- The viscosity of water and protoplasm is increased.
- Permeability of cell membranes is decreased.

- Metabolic activities of root cells are decreased.
- Growth and elongation of roots are checked.

Relative Importance of Active and Passive Absorption of Water

There are two views regarding the relative importance of active and passive absorption of water in the water economy of plants. Many workers in the past regarded the active absorption of water to be the main mechanism of water absorption and gave very little importance to the passive absorption. The active absorption of water is of negligible importance in the water economy of most or perhaps all plants.

He regards the root pressure and the related phenomena involved in the active absorption of water as mere consequences of salt accumulation in the xylem of different kinds of roots. The salt accumulation produces a difference in water potential which brings about the inward movement of water (osmotic uptake) and development of a pressure in the xylem sap (root pressure).

There are many reasons for regarding the active absorption as unimportant:

- The volume of exudates from the cut stump is very small in comparison to the volume of water lost in transpiration by the similar intact plants under conditions favorable for transpiration.
- Intact transpiring plants can absorb water from more concentrated and drier soil solutions more easily than the similar de-topped plants.
- No root pressure can be demonstrated in rapidly transpiring plants. Such plants may show even a negative root pressure (i.e., if a little water is placed over the cut stump it is absorbed by the latter).
- In conifers root pressure has rarely been observed.

It is held by certain workers that though the active absorption is not important quantitatively, it occurs all the time and supplements passive absorption. Two main arguments are against this view. Firstly, during periods of rapid transpiration the salts are removed from the root xylem so that their concentration becomes very low.

Under such conditions the osmotic uptake of water cannot be expected to occur. Secondly, even if we suppose that the salts are not removed during periods of rapid transpiration, the latter reduces the water potential of the cortical cells in roots to such a low level that the osmotic entry of water from cortex to xylem is not possible.

The available evidence suggests that usually the water is pulled passively into the plant through the roots by forces which are developed in the transpiring surfaces of the shoot. But under certain conditions such as warm moist soil and low rate of transpiration, salts accumulate in xylem of roots resulting in active osmotic absorption of water.

Field Capacity or Water Holding Capacity of the Soil

After heavy rainfall or irrigation of the soil, some water is drained off along the slopes while the rest percolates down in the soil. Out of this latter water some amount of water gradually reaches the water table under the force of gravity (gravitational water) while the rest is retained by the soil.

This amount of water retained by the soil after the drainage of gravitational water has become very slow is called as field capacity or the water holding capacity of the soil.

The field capacity is affected by soil profile, soil structure and temperature. For instance a fine textured soil overlying a coarse textured soil will have a higher field capacity than a uniformly fine textured soil. Similarly, the field capacity increases with decreasing temperature and vice versa.

Permanent Wilting Percentage or Wilting Coefficient

The percentage of the soil water left after the plant growing in that soil has permanently wilted is called as permanent wilting percentage or the wilting coefficient. The permanent wilting percentage can be determined by growing the seedlings in small containers under conditions of adequate water supply till they develop several leaves. The soil surface is then covered and the water supply is cut until wilting occurs. The containers are now transferred to humid chamber.

If the plants do not recover, they are considered to be permanently wilted. Otherwise, they are again transferred to normal atmospheric conditions. This process is repeated till they are permanently wilted. The percentage of the soil water is determined at this point after removing the plants from the containers and shaking off as much soil from their roots as possible.

Earlier workers thought permanent wilting percentage to be soil moisture constant. This view has been strongly criticized by Slatyer who pointed out that permanent wilting percentage of a soil is dependent on the osmotic characteristics of the plant and is not a soil-moisture constant. Thus the different plants if grown in the same soil wilt at different times depending upon their osmotic potential after the water supply to the soil is stopped.

Soil Texture in Relation to Water Absorption

The texture of a soil depends upon the proportion of different sized soil particles in that soil and is a very important factor for the absorption of water in plants.

Table: Depending upon their diameters the soil particles are classified as below.

Type of soil	Size
Gravel	2 mm. or more
Coarse sand	2—0.2 mm.
Fine sand	0.2—0.02 mm.
Silt	0.02—0.002 mm.
Clay	below 0.002 mm.

- Sandy Soils: Such soils are very rich in sand particles and though well aerated they have poor water holding capacity. Sandy soils are, therefore, not good for water absorption.
- Clayey Soils: These are rich in clay particles and are poorly aerated. Such soils often become water-logged and are, therefore, neither good for water absorption nor for normal growth of the plants.

- Loam: Such soils contain almost equal proportion of the different sized soil particles. They are sufficiently aerated and have good water holding capacity. Therefore, they are very good for water absorption and growth. The loam soil in which the proportion of sand is slightly higher is called as sandy loam while a loam soil in which clay particles predominate, is called as clayey loam.
- Velamen: Many epiphytic orchids develop special aerial adventitious roots which can absorb moisture from the atmosphere. For this purpose, a special water absorbing tissue is present around the cortex of such roots which is called as velamen. It consists of thin walled parenchymatous cells and the moisture absorbed by it is transferred to the root xylem through exodermis, cortex, endodermis and the pericycle.

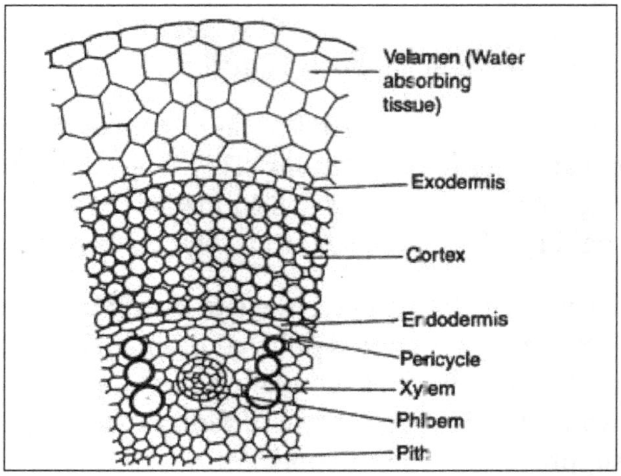

T.S. of a part of aerial root of an orchid showing velamen.

- Aquaporins: In recent years some integral membrane proteins have been discovered which form water selective channels in cell membranes (lipid bilayers) and facilitate faster movement of water across the membranes into the plant cells. These channels have been called as aquaporins. The direction of water transport across the membranes however, is not affected by aquaporins.

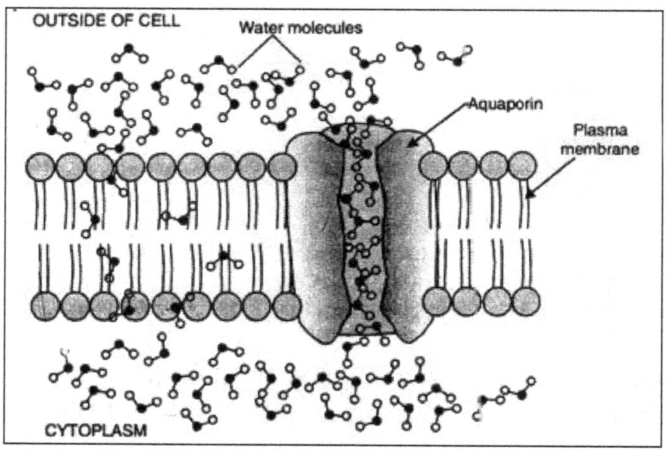

Bulk flow of water across the cell membrane through aquaporin.

Aquaporin's are found in both plant and animal membranes but they are relatively abundant in plants. The aquaporin's satisfactorily account for the observed rate of water movement across the membranes which could not be explained earlier simply by direct diffusion of water through lipid bilayer as the latter does not allow bulk flow of water across it.

Expression and activity of aquaporin's appear to be regulated probably by protein phosphorylation in response to availability of water.

Water Conduction

Vascular bundles in stem, roots, leaf veins forms an unbroken system of tubes which collectively perform a transport system throughout the entire plant body. Water and salts travel upwards through xylem tissues and food travels up and down directions through phloem tissues.

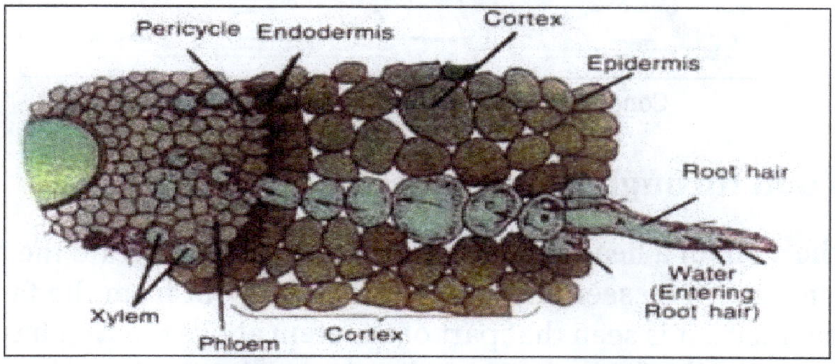

Cross section of a part of root showing cell-to-cell conduction of water from root hair to xylem.

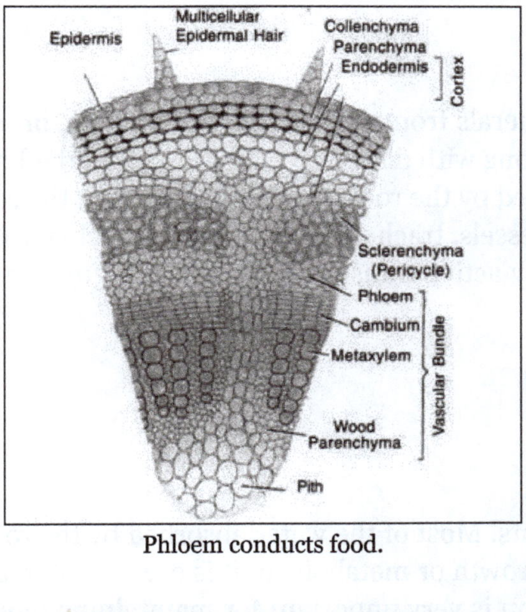

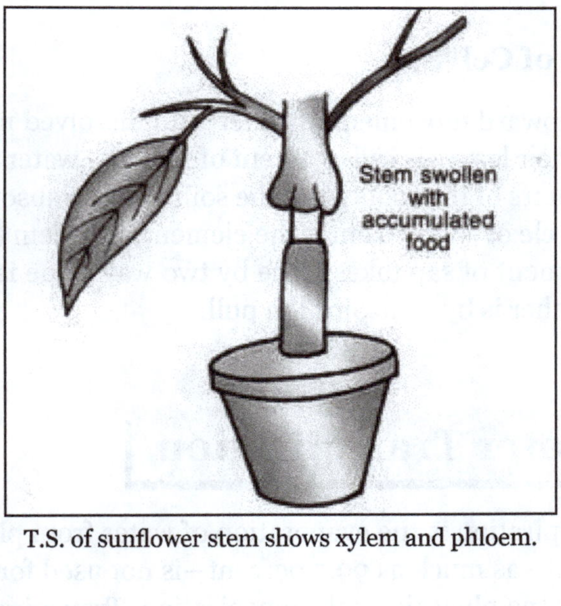

Phloem conducts food.

T.S. of sunflower stem shows xylem and phloem.

Conduction of Water through Xylem

Cut two leafy shoots of balsam plant under water to prevent entry of any air bubble. Now by keeping lower end of shoot in water remove near about 2-3 cm outer ring (phloem) of stem, keeping the central part intact. In second shoot, remove the central part (xylem) of equal length. Now fix both the shoots in stands and observe, after two days we find that the leaves of first shoot remain turgid and stand normally, but in second twig leaves wilt and droop down, this experiment proves that water is conducted through xylem.

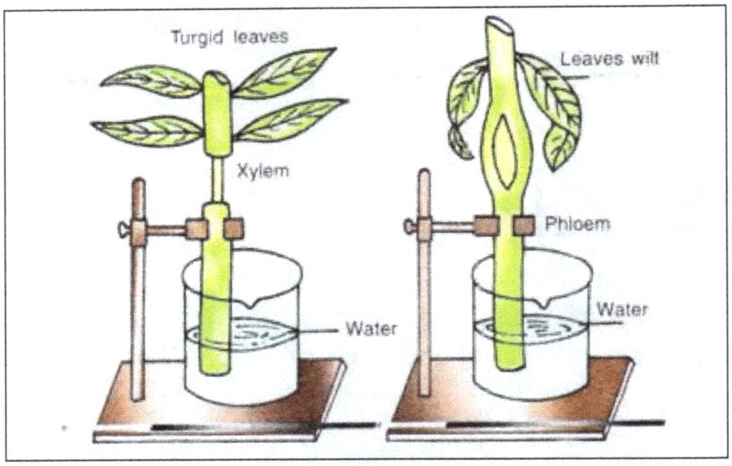

Conduction of ascent of sap through xylem.

Conduction of Food through Phloem

Cut a ring around the stem of a healthy plant deep enough to penetrate the phloem and cambium but not the xylem. It will be seen that sap starts oozing out from the farther cut-margin of the stem. After some weeks, it is seen that part of the stem above the ring has grown in diameter and below the ring the growth of stem is stopped and may die after some time, when food stored is finished. The leaves are healthy due to continuous supply of water through the xylem tissues.

Rise of Cell Sap

The upward movement of water with dissolved minerals from roots to the tips of stem, branches and their leaves is called ascent of sap. The water along with dissolved minerals is absorbed by the root hairs of the roots from the soil. The sap absorbed by the root hairs passes through the cortex, pericycle of roots to enter the elements of xylem (vessels, tracheids and xylem parenchyma). This movement of sap takes place by two ways, one is by active transportation which requires energy and other is by transpiration pull.

Water Transpiration

Transpiration is the evaporation of water from plants. Most of the water absorbed by the roots of a plant—as much as 99.5 percent—is not used for growth or metabolism; it is excess water, and it leaves the plant through transpiration. Transpiration is very important for maintaining moisture conditions in the environment. As much as 10 percent of the moisture in the Earth's atmosphere is from transpiration of water by plants.

Function of Transpiration

Transpiration occurs because plants take in more water than they actually need at a given time. It is a way of getting rid of excess water. When water is removed from the plant, it can more easily access the carbon dioxide that it needs for photosynthesis. Also, plants can use transpiration as a method of cooling themselves.

Transpiration is used to describe the specific action of water evaporating from a plant, but the word transpiration is also used to generally describe how water moves through plants. When water enters the plant through the roots, it is pulled up through the xylem tissue in the stem of the plant to the plant's leaves by capillary action and the cohesion of water molecules. When water reaches the stomata, which are small holes in the leaves, it evaporates due to diffusion; the moisture content of the air is lower than the moisture in the leaf, so water naturally flows out into the surrounding air in order to equalize the concentrations.

Transpiration has side effects for other organisms in an ecosystem. It helps maintain a certain moisture level in an environment, depending on the number and types of plants in an environment. This inadvertently allows some organisms to survive better than others depending on the moisture levels that they need to thrive.

Examples of Transpiration

- Stomatal Transpiration: Stomatal transpiration is the evaporation of water from a plant's stomata. Most of the water that is transpired from a plant is transpired this way; at least 90% of the water transpired from a plant's leaves exits through the stomata. Near the surface of the leaf, water in liquid form changes to water vapor and evaporates from the plant through open stomata.
- Cuticular Transpiration: Cuticular transpiration is the evaporation of water from a plant's cuticle. The cuticle is a waxy film that covers the surface of a plant's leaves. This form of transpiration does not account for much of a plant's water loss; about 5-10 percent of the leaves' water is lost through the cuticle. When plants close their stomata in dry conditions, more water is transpired this way.
- Lenticular Transpiration: Lenticular transpiration is the evaporation of water from the lenticels of a plant. Lenticels are small openings in the bark of branches and twigs. Not all plants have lenticels. The amount of water lost this way is very small compared to stomatal transpiration, but as with cuticular transpiration, it may increase if a plant is in a dry environment.

Factors that Affect Transpiration

There are many factors that affect transpiration. One such factor is temperature. When temperatures increase, the stomata of leaves open and more water transpires. Plants that grow in warmer climates transpire more. Moisture levels of the air and soil are other important factors. When relative humidity of the air increases, there is more moisture in the air, so transpiration decreases. However, if there is more moisture in the soil, plants will transpire more because they are taking in more water. More wind also increases the rate of transpiration because it decreases the relative humidity around a plant. Of course, some plants also just transpire more than others. Plants that live in dry environments, such as cacti, have evolved to conserve water in part by transpiring less water. This allows them to thrive in arid regions like the desert.

The Water Cycle

Transpiration is part of the water cycle, also known as the hydrological cycle. The water cycle describes how water moves throughout the Earth. First, water transpires from plants and enters the atmosphere as water vapor. Water from Earth's oceans, lakes, and rivers also evaporates into the atmosphere. The evaporation from Earth's waterways and from plants via transpiration is collectively known as evapotranspiration. In the atmosphere, the water forms clouds, and then it falls back to earth again as rain or snow. Precipitation collects again in earth's waterways, or it goes into the soil, where it enables plants to grow. Then water evaporates from plants, oceans, lakes, and rivers again, completing the cycle.

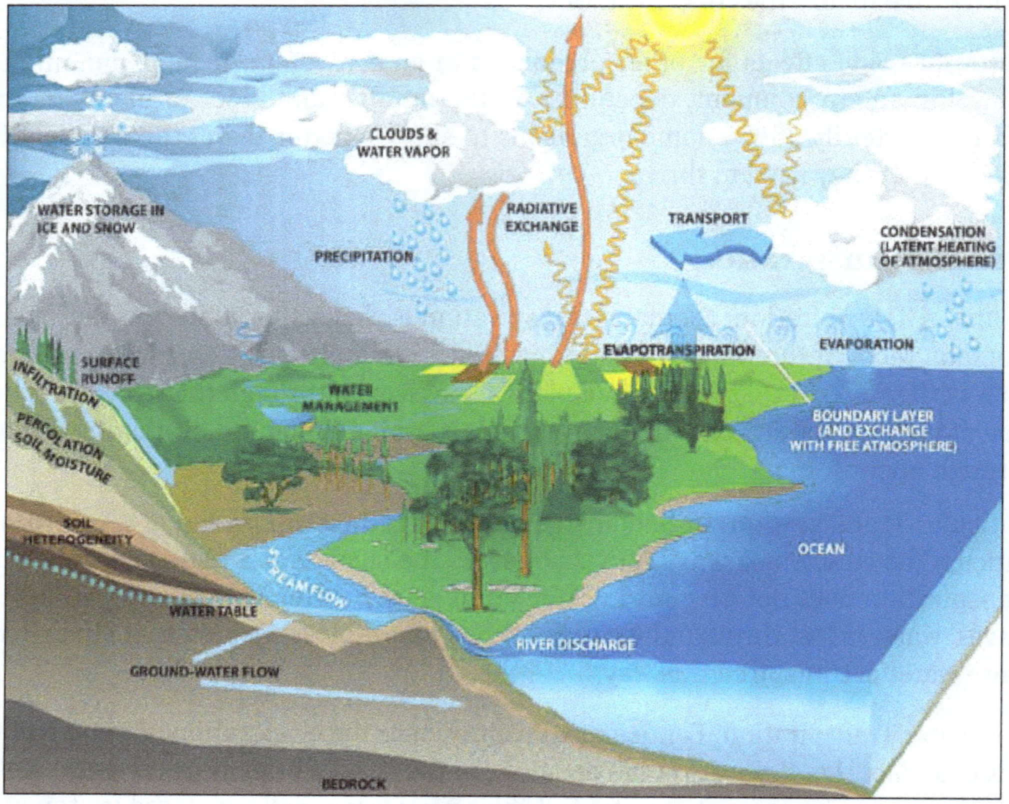

This diagram shows the Earth's water cycle.

References

- Plant-water-relations, biology: byjus.com, Retrieved 16, March 2020
- Difference-between-endosmosis-and-exosmosis: pediaa.com, Retrieved 21, May 2020
- Defining-water-potential, environment: metergroup.com, Retrieved 12, July 2020
- Water-uptake-and-transport-in-vascular-plants-103016037: nature.com, Retrieved 19, June 2020
- Absorption-of-water-in-plants-with-diagram-22718: biologydiscussion.com, Retrieved 09, Feb 2020

Chapter 4

Irrigation: Techniques and Methods

There are several methods of irrigation like surface irrigation, micro-irrigation, drip irrigation, sprinkler irrigation, subsurface textile irrigation, center pivot irrigation, garden hose, subirrigation etc. Surface irrigation is the oldest form of irrigation. In this method, water moves across the surface of fields in order to infiltrate into the soil. Various techniques and methods of irrigation have been thoroughly discussed in this chapter.

Surface Irrigation

Surface irrigation has evolved into an extensive array of configurations, which can be broadly classified as:

(1) Basin irrigation.

(2) Border irrigation.

(3) Furrow irrigation.

(4) Uncontrolled flooding.

There are two features that distinguish a surface irrigation system:

(a) The flow has a free surface responding to the gravitational gradient.

(b) The on-field means of conveyance and distribution is the field surface itself.

A surface irrigation event is composed of four phases as illustrated graphically in figure below. When water is applied to the field, it 'advances' across the surface until the water extends over the entire area. It may or may not directly wet the entire surface, but all of the flow paths have been completed. Then the irrigation water either runs off the field or begins to pond on its surface. The interval between the end of the advance and when the inflow is cut off is called the wetting or ponding phase. The volume of water on the surface begins to decline after the water is no longer being applied. It either drains from the surface (runoff) or infiltrates into the soil. For the purposes of describing the hydraulics of the surface flows, the drainage period is segregated into the depletion phase (vertical recession) and the recession phase (horizontal recession). Depletion is the interval between cut off and the appearance of the first bare soil under the water. Recession begins at that point and continues until the surface is drained.

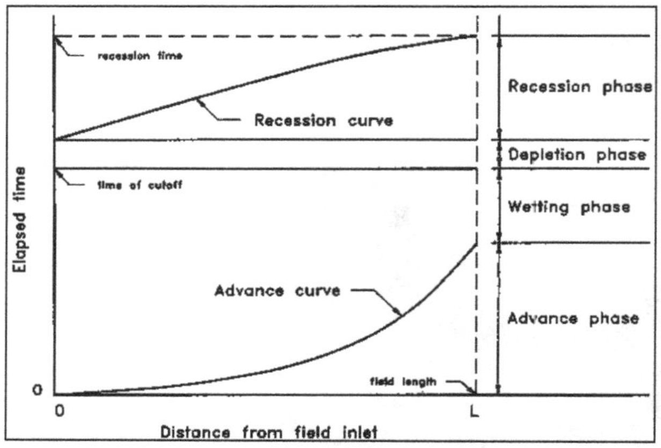

Time-space trajectory of water during a surface irrigation showing its advance, wetting, and depletion and recession phases.

The time and space references shown in figure above are relatively standard. Time is cumulative since the beginning of the irrigation, distance is referenced to the point water enters the field. The advance and recession curves are therefore trajectories of the leading and receding edges of the surface flows and the period defined between the two curves at any distance is the time water is on the surface and therefore also the time water is infiltrating into the soil.

It is useful to note here that in observing surface irrigation one may not always observe a ponding, depletion or recession phase. In basins, for example, the post-cut off period may only involve a depletion phase as the water infiltrates vertically over the entire field. Likewise, in the irrigation of paddy rice, irrigation very often adds to the ponded water in the basin so there is neither advance nor recession - only wetting or ponding phase and part of the depletion phase. In furrow systems, the volume of water in the furrow is very often a small part of the total supply for the field and it drains rapidly. For practical purposes, there may not be a depletion phase and recession can be ignored. Thus, surface irrigation may appear in several configurations and operate under several regimes.

The surface irrigation system is one component of a much larger network of facilities diverting and delivering water to farmlands. Figure above illustrates the 'irrigation system' and some of its features. It may be divided into the following four component systems:

- Water supply.
- Water conveyance or delivery.
- Water use.
- Drainage.

For the complete system to work well, each must work conjunctively toward the common goal of promoting maximum on-farm production. Historically, the elements of an irrigation system have not functioned well as a system and the result has too often been very low project irrigation efficiencies.

The focus of surface irrigation engineering is at the water use level, the individual irrigated field. For design and evaluation purposes, these guidelines will note elements of the conveyance and distribution system, especially those near the field such as flow measurement and control, but will leave detailed treatment to other technical sources.

Evolution of the Practice

Although surface irrigation is thousands of years old, the most significant advances have been made within the last decade. In the developed and industrialized countries, land holdings have become as much as 10-20 times as large, and the number of farm families has dropped sharply. Very large mechanized farming equipment has replaced animal-powered planting, cultivating and harvesting operations. The precision of preparing the field for planting has improved by an order of magnitude with the advent of the laser-controlled land grading equipment. Similarly, the irrigation works themselves are better constructed because of the application of high technology equipment.

The changes in the lesser-developed and developing countries are less dramatic. In the lesser-developed countries, trends toward land consolidation, mechanization, and more elaborate system design and operation are much less apparent. Most of these farmers own and operate farms of 1-10 hectares, irrigate with 20-40 liters per second and rely on either small mechanized equipment or animal-powered farming implements.

Probably the most interesting evolution in surface irrigation so far as this guide is concerned is the development and application of microcomputers and programmable calculators to the design and operation of surface irrigation systems. In the late 1970s, a high-speed microcomputer technology began to emerge that could solve the basic equations describing the overland flow of water quickly and inexpensively. At about the same time, researchers like Strelkoff and Katapodes made major contributions with efficient and accurate numerical solutions to these equations. Today in the graduate and undergraduate study of surface irrigation engineering, microcomputer and programmable calculator utilization is, or should be, common practice.

Microcomputers and programmable calculators provide several features for today's irrigation engineers and technicians. They allow a much more comprehensive treatment of the vital hydraulic processes occurring both on the surface and beneath it. One can find optimal designs and management practices for a multitude of conditions because designs historically requiring days of effort are now made in seconds. The effectiveness of existing practices or proposed ones can be predicted, even to the extent that control systems operating, sensing and adjusting on a real-time basis are possible.

Surface Irrigation Methods

The classification of surface methods is perhaps somewhat arbitrary in technical literature. This has been compounded by the fact that a single method is often referred to with different names. In this guide, surface methods are classified by the slope, the size and shape of the field, the end conditions, and how water flows into and over the field.

Each surface system has unique advantages and disadvantages depending on such factors as were listed earlier like:

- Initial cost.
- Size and shape of fields.
- Soil characteristics.
- Nature and availability of the water supply.
- Climate.
- Cropping patterns.
- Social preferences and structures.
- Historical experiences.
- Influences external to the surface irrigation system.

Basin Irrigation

Basin irrigation is the most common form of surface irrigation, particularly in regions with layouts of small fields. If a field is level in all directions, is encompassed by a dyke to prevent runoff, and provides an undirected flow of water onto the field, it is herein called a basin. A basin is typically square in shape but exists in all sorts of irregular and rectangular configurations. It may be furrowed or corrugated, have raised beds for the benefit of certain crops, but as long as the inflow is undirected and uncontrolled into these field modifications, it remains a basin.

There are few crops and soils not amenable to basin irrigation, but it is generally favored by moderate to slow intake soils, deep-rooted and closely spaced crops. Crops, which are sensitive to flooding, and soils which form a hard crust following an irrigation can be basin irrigated by adding furrowing or using raised bed planting. Reclamation of salt-affected soils is easily accomplished with basin irrigation and provision for drainage of surface runoff is unnecessary. Of course it is always possible to encounter a heavy rainfall or mistake the cut-off time thereby having too much water in the basin. Consequently, some means of emergency surface drainage is good design practice. Basins can be served with less command area and field watercourses than can border and furrow systems because their level nature allows water applications from anywhere along the basin perimeter. Automation is easily applied.

Basin irrigation has a number of limitations, two of which, already mentioned, are associated with soil crusting and crops that cannot accommodate inundation. Precision land leveling is very important to achieving high uniformities and efficiencies. Many basins are so small that precision equipment cannot work effectively. The perimeter dykes need to be well maintained to eliminate breaching and waste, and must be higher for basins than other surface irrigation methods. To reach maximum levels of efficiency, the flow per unit width must be as high as possible without causing erosion of the soil. When an irrigation project has been designed for either small basins or furrows and borders, the capacity of control and outlet structures may not be large enough to improve basins.

Border Irrigation

Border irrigation can be viewed as an extension of basin irrigation to sloping, long rectangular or contoured field shapes, with free draining conditions at the lower end. Water is applied to individual borders from small hand-dug checks from the field head ditch. When the water is shut off, it recedes from the upper end to the lower end. Sloping borders are suitable for nearly any crop except those that require prolonged ponding. Soils can be efficiently irrigated which have moderately low to moderately high intake rates but, as with basins, should not form dense crusts unless provisions are made to furrow or construct raised borders for the crops. The stream size per unit width must be large, particularly following a major tillage operation, although not so large for basins owing to the effects of slope. The precision of the field topography is also critical, but the extended lengths permit better leveling through the use of farm machinery.

Furrow Irrigation

Furrow irrigation avoids flooding the entire field surface by channeling the flow along the primary direction of the field using 'furrows,' 'creases,' or 'corrugations'. Water infiltrates through the wetted perimeter and spreads vertically and horizontally to refill the soil reservoir. Furrows are often employed in basins and borders to reduce the effects of topographical variation and crusting. The distinctive feature of furrow irrigation is that the flow into each furrow is independently set and controlled as opposed to furrowed borders and basins where the flow is set and controlled on a border by border or basin by basin basis.

Furrows provide better on-farm water management flexibility under many surface irrigation conditions. The discharge per unit width of the field is substantially reduced and topographical variations can be more severe. A smaller wetted area reduces evaporation losses. Furrows provide the irrigator more opportunity to manage irrigations toward higher efficiencies as field conditions change for each irrigation throughout a season. This is not to say, however, that furrow irrigation enjoys higher application efficiencies than borders and basins.

There are several disadvantages with furrow irrigation. These may include:

- An accumulation of salinity between furrows.
- An increased level of tailwater losses.
- The difficulty of moving farm equipment across the furrows.
- The added expense and time to make extra tillage practice (furrow construction).
- An increase in the erosive potential of the flow.
- A higher commitment of labor to operate efficiently.
- Generally furrow systems are more difficult to automate, particularly with regard to regulating an equal discharge in each furrow.

(a) Graded furrow irrigation system:

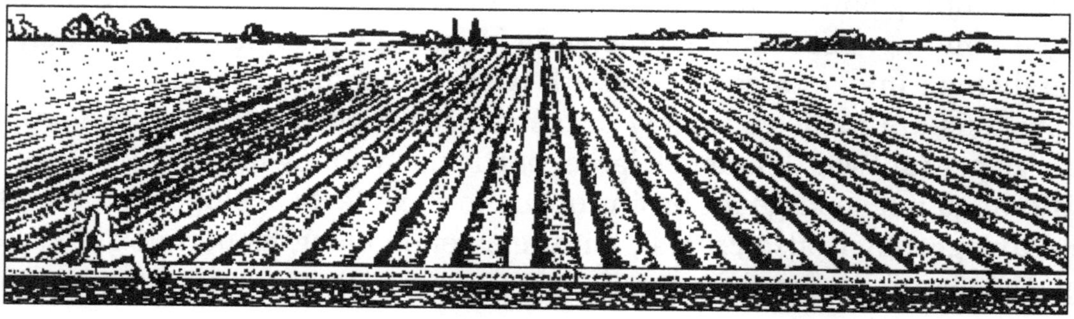

(b) Contour furrows:

Uncontrolled Flooding

There are many cases where croplands are irrigated without regard to efficiency or uniformity. These are generally situations where the value of the crop is very small or the field is used for grazing or recreation purposes. Small land holdings are generally not subject to the array of surface irrigation practices of the large commercial farming systems. Also in this category are the surface irrigation systems like check-basins which irrigate individual trees in an orchard, for example. The evaluation methods can be applied if desired, but the design techniques are not generally applicable nor need they be since the irrigation practices tend to be minimally managed.

Requirements for Optimal Performance

There is substantial field evidence that surface irrigation systems can apply water to croplands uniformly and efficiently, but it is the general observation that most such systems operate well below their potential. A very large number of causes of poor surface irrigation performance have been outlined in the technical literature. They range from inadequate design and management at the farm level to inadequate operation of the upstream water supply facilities. However, in looking for a root cause, one most often retreats to the fact that infiltration changes a great deal from irrigation to irrigation, from soil to soil, and is neither predictable nor effectively manageable. The infiltration rates are an unknown variable in irrigation practice.

In those cases where high levels of uniformity and efficiency are being achieved, irrigators utilize one or more of the following practices:

- Precise and careful field preparation.
- Irrigation scheduling.
- Regulation of inflow discharges.
- Tail water runoff restrictions, reduction, or reuse.

Inlet Discharge Control

Surface irrigation systems have two principal sources of inefficiency, deep percolation and surface runoff or tail water. The remedies are competitive. To minimize deep percolation the advance phase should be completed as quickly as possible so that the intake opportunity time over the field will be uniform and then cut the inflow off when enough water has been added to refill the root zone. This can be accomplished with a high, but non-erosive, discharge onto the field. However, this practice increases the tail water problem because the flow at the downstream end must be maintained until a sufficient depth has infiltrated. The higher inflow reaches the end of the field sooner but it increases both the duration and the magnitude of the runoff.

There are three options available to solve this problem, at least partially:

- Dyke the downstream end to prevent runoff as in basin irrigation.
- Reduce the inflow discharge to a rate more closely approximating the cumulative infiltration along the field following the advance phase, a practice termed 'cutback'.
- Select a discharge, which minimizes the sum of deep percolation and tail water losses, i.e., optimize the field inflow regime.

In this configuration, the head ditch is divided into a series of level bays, which are differentiated by a small change in elevation. Water levels are regulated in two bays simultaneously so that the lower bay has sufficient head to produce an advance phase flow in the furrows while in the upper bay the head is only sufficient to produce the cutback flow. Thus, the system operates by moving the check-dam from bay to bay along the upper end of the field.

Two very recent additions to the efforts to control surface irrigation systems more effectively are the 'Surge Flow' system developed at Utah State University, USA and the 'Cablegation' system developed at the US Department of Agriculture's Snake River Water Conservation Research Center in Kimberly, Idaho, USA.

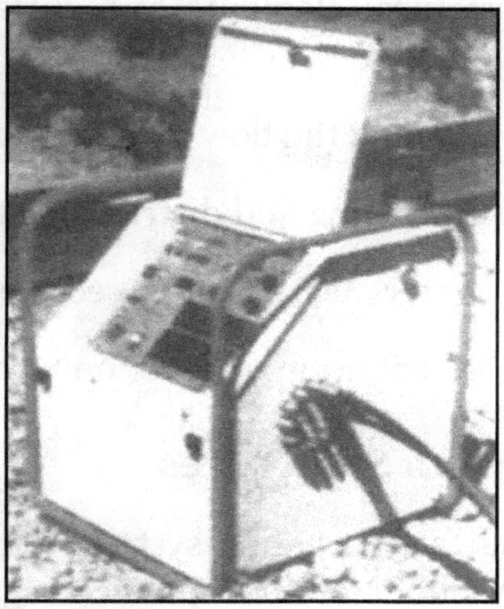

Wastewater Recovery and Reuse

The tail water deep percolation trade-off can also be solved by collecting and recycling the runoff to improve surface irrigation performance. Reuse systems have not been widely employed historically because water and energy have been inexpensive. Even today it is often more economical to regulate the inflow rather than to collect and pump the runoff back to the head of the field or to another field, tail water reuse systems are more cost-effective when the water can be added to the flow serving lower fields and thereby saving the cost of pumping.

Surface Irrigation Structures

Surface irrigation systems are supported by a number of on- and off-farm structures which control and manage the flow and its energy. In order to facilitate efficient surface irrigation, these structures should be easily and cheaply constructed as well as easy to manage and maintain. Each should be standardized for mass production and fabrication in the field by farmers and technicians.

It is not the intent of this guide to be comprehensive with regard to the selection and design of these structures since other sources are available, but it is worthwhile to note some of these structures by way of presenting a larger view of surface irrigation.

The structural elements of a surface system perform several important functions, which include:

- Turning the flow to a field on and off.
- Conveying and distributing the flow among fields.
- Water measurement, sediment and debris removal, and water level stabilization.
- Distribution of water onto the field.

Diversion Structures

Most surface irrigation systems derive their water supplies from canal systems operated by public or semi-public irrigation departments, districts, or companies. Some irrigation water is supplied in piped delivery systems and some directly pumped from groundwater. Diversion structures perform several tasks including:

- On-off water control, which allows the supply agency to allocate its supply, and protects the fields below the diversion from untimely flooding.
- Regulation and stabilization of the discharge to the requirements of field channels and watercourse distribution systems.
- Measurement of flow at the turnout in order to establish and protect water entitlements.
- Protection of downstream structures by controlling sediments and debris as well as dissipating excess kinetic energy in the flow. A typical turnout structure is shown in figure below.

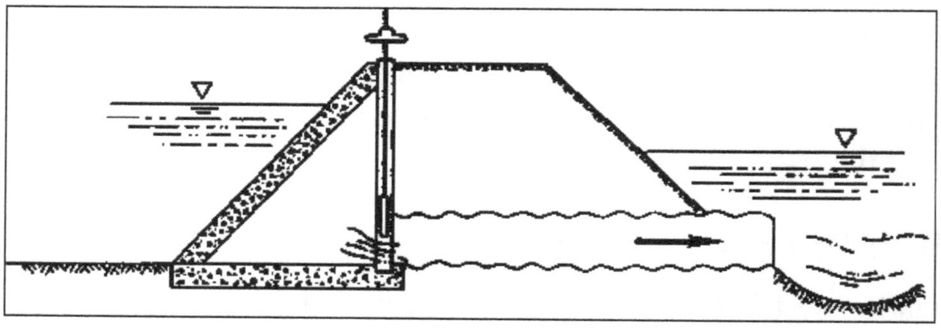

Typical turnout from a canal or lateral.

Conveyance, Distribution and Management Structures

Conveying water to the field requires similar structures to those found in major canal networks. The conveyance itself can be an earthen ditch or lateral, a buried pipe, or a lined ditch. Lined sections can be elevated or constructed at surface level. Pipe materials are usually plastic, steel, concrete, clay, or asbestos cement, or they may be as simple as a wooden or bamboo construction. Lining materials include slip-form cast-in-place, or prefabricated concrete, shotcrete or gunite, asphalt, surface and buried plastic or rubber membranes, and compacted earth.

The management of water in the field channels involves flow measurement, sediment and debris removal, divisions, checks, drop-energy dissipators, and water level regulators. Some of the more common flow control structures for open channels are shown in figure below. Associated with these are various flow measuring devices like weirs, flumes, and orifices. The designs of these structures have been standardized since they are small in size and capacity. Designs for flow measurement and drop-energy dissipator structures need more attention and construction must be more precise since their hydraulic responses are quite sensitive to their dimensions.

The figure below shows on-farm water management structures;

A simple drop structure.

A typical check-divider.

Field Distribution Systems

After the water reaches the field ready to be irrigated, it is distributed onto the field by a variety of means, both simple and elaborately constructed. Most fields have a head ditch or pipeline running along the upper side of the field from which the flow is distributed onto the field.

In a field irrigated from a head ditch, the spreading of water over the field depends somewhat on the method of surface irrigation. For borders and basins, open or piped cutlets are generally used. Furrow systems use outlets which can be directed to each furrow.

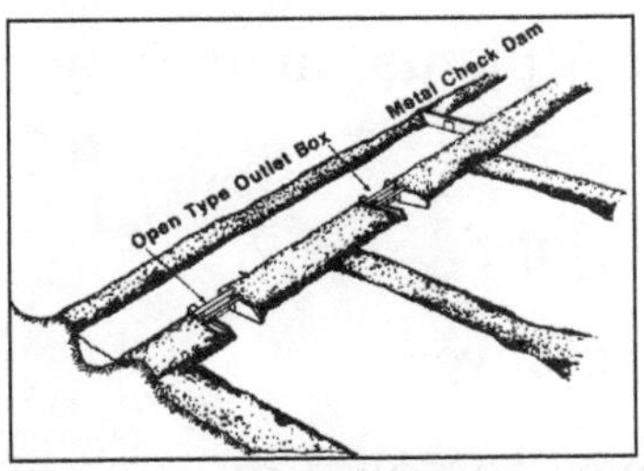

Head ditch outlets for borders and basins.

Field distribution and spreading can also be through portable pipelines running along the surfaces or permanent pipelines running underground. Basins and borders usually receive water through buried pipes serving one or more gated risers within each basin or border. A typical riser outlet, known as an alfalfa valve, is shown in figure. The most common piped method of furrow irrigation uses plastic or aluminium gated pipe like that shown in figure. The gated pipe may be connected to the main water supply via a piped distribution network with a riser assembly like the one shown in figure, directly to a canal turnout, or through an open channel to a piped transition.

An alfalfa valve riser.

Benefits

- Low cost of deployment, energy and maintenance.
- It favors the increase of the photosynthesis in the lower leaves, due to the reflection of the light in the water.
- The wind does not limit the irrigation.
- It promotes the fixation of atmospheric nitrogen, as a result of favoring the growth of blue-green algae.

Disadvantages

- Standing water can harm plants, mainly by reducing the respiration of the roots.
- Fairly dependent on soil slope.
- Erosions frequent in the grooves.
- Loss of water occurs due to percolation.

Micro Irrigation

Conventional irrigation systems, such as channel irrigation and wild flooding tend to waste water as large quantities are supplied to the field in one go, most of which just flows over the crop and runs away without being taken up by the plants.

Micro irrigation is an approach to irrigation that keeps the water demand to a minimum. It has been driven by commercial farmers in arid regions of the United States of America and Israel in farming areas where water is scarce.

Typically, these commercial irrigation systems consist of a surface or buried pipe distribution network using emitters supplying water directly to the soil at regular intervals along the pipework. They can be permanent or portable.

Kenya Practical Action staff demonstrate drip irrigation.

Many parts of the world are now using micro irrigation technology. The systems used by large commercial companies are generally quite complex with an emphasis on reducing the amount of labor involved.

Small-scale farmers in developing countries have been reluctant to take up micro irrigation methods due to the initial investment required for the equipment. A number of organizations have looked at ways to simplify and reduce the cost of micro irrigation resulting in the approaches of drip irrigation and pipe irrigation. For these small-scale irrigation systems not only should the technical aspects of the system be considered such as:

- Access to reliable water sources.
- A secure and well-fenced garden.
- Basic gardening skills.
- The crops grown.

The social aspects should also be of concern to ensure the irrigation system will be of benefit. The social and economical factors will include:

- Capital and financial management; credit facilities.
- The availability of external services.
- Maintenance.
- Market opportunities for the produce.
- Willingness to show other farmers the technology.

Drip Irrigation

Drip irrigation uses low-cost plastic pipes laid on the ground to irrigate vegetables, field crops and orchards. This technology was developed in the 1960s for commercial use. Circa 1990 a US firm called Chapin Watermatics developed a low-cost system called bucket kits, which use standard plastic buckets and lengths of hose that could be cut to the appropriate lengths.

Small holes in the hose allow water to drip out and keep the base of the plant wet without wasting any water. The kits are low-cost, easy to assemble and manage. They do not need high quality water, providing the water is filtered. A 20 liter bucket with 30 meters (100 feet) of hose or drip tape connected to the bottom. The bucket is placed at least 1 meter (3 feet) above the ground so that gravity provides sufficient water pressure to ensure even watering for the entire crop. Water is poured into the bucket twice daily and passes through a filter, fills the drip tape and is evenly distributed to 100 watering points. The multi-chambered plastic drip tape is engineered to dispense water through openings spaced at 30cm (12 inches).

Two bucket kits costing around $20 will produce enough vegetables for a family of seven and can last over five years. The system is most suited to kitchen gardens. As well as the bucket, you will need several strong poles, tools, manure, water and vegetable seedlings. The poles are used to make a support structure for the bucket. The stand should hold the bucket about 1 meter above the ground. The main stages of setting up the systems are:

Practical Action farmers project staff examine the drip irrigation kit.

- A hole should be cut carefully into the base of the bucket.
- The hole is fitted with the filter plug and tubing and then flushed out to ensure that the system is clean.
- The drip lines are then connected and the system is flushed out again before the ends of the drip lines are closed off.

The whole procedure of setting up the system will only take about one hour, including the construction of the bucket support. When planting a seedling is planted at each wet spot so that all the moisture is absorbed directly by the plant roots. Moving the kit from plot to plot tends not to be very practical and damages the equipment. It is better to add extra buckets and lines when necessary or when funds are available to invest in additional equipment.

The advantages are:

- The effort to water the plants is greatly reduced.
- The time taken to fill the containers is significantly less than manually watering the planted area.
- The growth of weeds is reduced, as water does not reach unwanted plants.

Some initial training is required to show how to get the best results from the system and careful attention should be taken to a number of common problems that occur with the system comprising of:

- Clogging of the trip tapes especially if water is not well filtered. The emitters can be cleaned by blowing dirt out and then flushing out the particles with clean water.
- Leakage at connections: this can be checked and corrected.

- Weeding must avoid puncturing the tape.
- Vermin can damage pipes in their search for water.
- Theft of equipment.

The bucket kit is the smallest type of drip irrigation available. Although productivity is increased providing greater food security and improved nutrition, the amount of water and labour saved is small. Often there is not sufficient surplus produce for users to sell and provide financial benefits.

Bigger containers can be used to suit larger market gardens. Large customized drum kits irrigating a high-value crop offer greatest financial impact.

In East Africa drip irrigation has been promoted by Practical Action East Africa and the Arid Land Information Network (ALIN) who sell drip irrigation kits similar to the ones used in India and Zimbabwe.

Water is poured into the pipe which then irrigates the plants at the roots.

Pipe Irrigation

The use of buried clay pitchers is an ancient technique of subsurface irrigation. The use of clay pipes had been initially tried in Russia and Mexico. ITDG Southern Africa along with others in Zimbabwe developed a low cost variation of this irrigation method in which clay pipes were buried beneath vegetable beds of the 450 women members of garden groups.

One end of the pipe is blocked and the other is tilted out of the soil to allow filling. The pipes were laid end to end below the soil surface. When the pipe is filled, water will gradually escape from the cracks between the pipe sections and through the pores in the clay to provide a continuous supply of water to the vegetables.

There were difficulties in obtaining the correct dimensions. Research at the Lowveld Research Station, Zimbabwe found that the optimum pipe size was 75mm inside diameter and 300mm long.

The pipes are produced using a mould made from a plastic drainpipe with handles attached. The clay is prepared by rolling it out to a suitable size and thickness before it is put into the mould. Once it has been fitted to the mould the pipe can be removed and stood on end to dry naturally. The pipes are fired in an open fire.

Although the pipes are still laborious to fill by bucket, the use of clay pipes reduced the work for the women because the vegetable gardens needed to be watered only once a week instead of 3-4 times. Watering was said to be reduced by 50% and vegetable production increased.

The main benefit of the clay pipe approach is that the pipes were manufactured by the women, who ran the vegetable garden, without any external input. The disadvantages are that it is a long and laborious task to produce, that the installation of clay pipes has to be carried out every season, and the pipes are easily broken during garden operations.

Need of Micro Irrigation

- To make agriculture productive,
- Environmentally sensitive and capable of preserving the social fabric of rural communities,
- Help produce more from the available land, water and labor resources without either ecological or social harmony,
- Generate higher farm income,
- On-farm and off-farm employment.

Drip Irrigation

Drip irrigation is sometimes called trickle irrigation and involves dripping water onto the soil at very low rates (2-20 liters/hour) from a system of small diameter plastic pipes fitted with outlets called emitters or drippers. Water is applied close to plants so that only part of the soil in which the roots grow is wetted, unlike surface and sprinkler irrigation, which involves wetting the whole soil profile. With drip irrigation water, applications are more frequent (usually every 1-3 days) than with other methods and this provides a very favorable high moisture level in the soil in which plants can flourish.

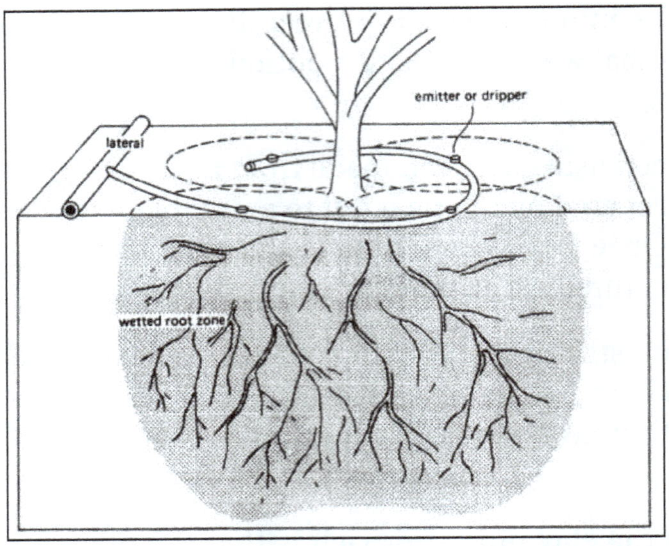

Drip irrigation system delivers water to the crop using a network of mainlines, submains and lateral lines with emission points spaced along their lengths. Each dripper/emitter, orifice supplies a measured, precisely controlled uniform application of water, nutrients, and other required growth substances directly into the root zone of the plant.

Water and nutrients enter the soil from the emitters, moving into the root zone of the plants through the combined forces of gravity and capillary. In this way, the plant's withdrawal of moisture and nutrients are replenished almost immediately, ensuring that the plant never suffers from water stress, thus enhancing quality, its ability to achieve optimum growth and high yield.

Drip System Layout

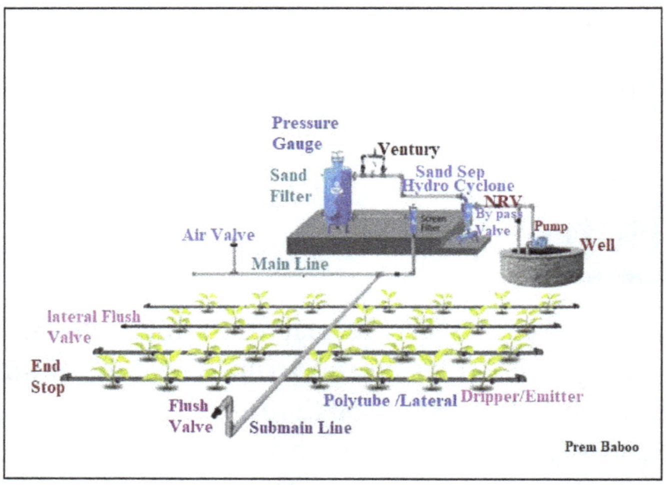

Major Components of Drip Irrigation System			
1	Pump station	2	By-pass assembly
3	Control valves	4	Filtration system

5	Fertilizer tank /Venturi	6	Pressure gauge
7	Mains / Sub-mains	8	Laterals
9	Emitting devices	10	Micro tubes

- Pump station takes water from the source and provides the right pressure for delivery into the pipe system.
- Control valves control the discharge and pressure in the entire system.
- Filtration system cleans the water. Common types of filter include screen filters and graded sand filters, which remove fine material suspended in the water.
- Fertilizer tank/venturi slowly add a measured dose of fertilizer into the water during irrigation. This is one of the major advantages of drip irrigation over other methods.
- Mainlines, submains and laterals supply water from the control head into the fields. They are usually made from PVC or polyethylene hose and should be buried below ground because they easily degrade when exposed to direct solar radiation. Lateral pipes are usually 13-32 mm diameter.
- Emitters or drippers are devices used to control the discharge of water from the lateral to the plants. They are usually spaced more than 1 meter apart with one or more emitters used for a single plant such as a tree. For row crops more closely spaced emitters may be used to wet a strip of soil. Many different emitter designs have been produced in recent years. The basis of design is to produce an emitter, which will provide a specified constant discharge which does not vary much with pressure changes, and does not block easily.

Wetting Pattern in Drip Irrigation

Unlike surface and sprinkler irrigation, drip irrigation only wets part of the soil root zone. This may be as, low as 30% of the volume of soil wetted by the other methods. The wetting patterns which develop from dripping water onto the soil depend on discharge and soil type. Figure below shows the effect of changes in discharge on two different soil types, namely sand and clay.

Although only part of the root zone is wetted it is still important to meet the full water needs of the crop. It is sometimes thought that drip irrigation saves water by reducing the amount used by the crop. This is not true. Crop water use is not changed by the method of applying water. Crops just require the right amount for good growth.

The water savings that can be made using drip irrigation are the reductions in deep percolation, in surface runoff and in evaporation from the soil. These savings, it must be remembered, depend as much on the user of the equipment as on the equipment itself.

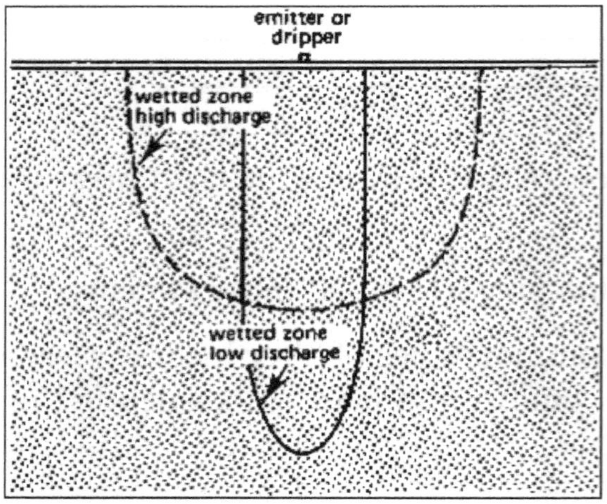

Wetting pattern in Sandy Soils.

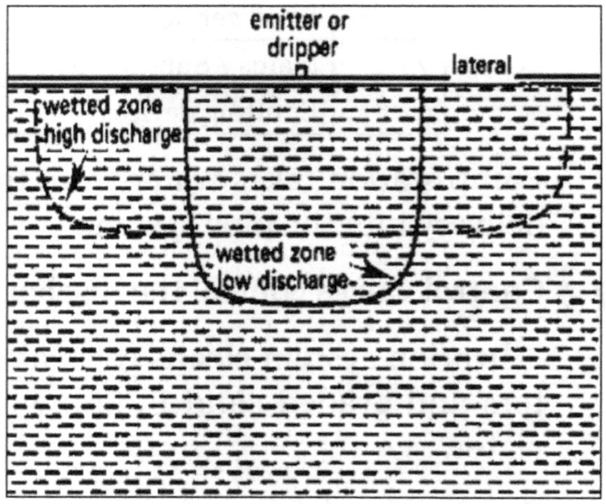

Wetting pattern in Clay Soils.

Drip irrigation is not a substitute for other proven methods of irrigation. It is just another way of applying water. It is best suited to areas where water quality is marginal, land is steeply sloping or undulating and of poor quality, where water or labor are expensive, or where high value crops require frequent water applications.

Crops Suitable for Drip Irrigation System

1	Orchard Crops	Grapes, Banana, Pomegranate, Orange, Citrus, Mango, Lemon, Custard Apple, Sapota, Guava, Pineapple, Coconut, Cashewnut, Papaya, Aonla, Litchi, Watermelon, Muskmelon etc.
2	Vegetables	Tomato, Chilly, Capsicum, Cabbage, Cauliflower, Onion, Okra, Brinjal, Bitter Gourd, Ridge Gourd, Cucumber, Peas, Spinach, Pumpkin etc.
3	Cash Crops	Sugarcane, Cotton. Arecanut, Strawberry etc.
4	Flowers	Rose, Carnation, Gerbera, Anthurium, Orchids, Jasmine, Dahilia, Marigold etc.
5	Plantation	Tea, Rubber, Coffee, Coconut etc.
6	Spices	Turmeric, Cloves, Mint etc,
7	Oil Seed	Sunflower, Oil palm, Groundnut etc.
8	Forest Crops	Teakwood, Bamboo etc.

Response of Different Crops to Drip Irrigation System

Crops	Water saving (%)	Increase in yield (%)
Banana	45	52
Cauliflower	68	70
Chilly	68	28
Cucumber	56	48
Grapes	48	23
Ground nut	40	152

Pomegranate	45	45
Sugarcane	50	99
Sweet lime	61	50
Tomato	42	60
Watermelon	66	19

Benefits of Drip Irrigation

- Increase in yield up to 230 %.
- Saves water up to 70% compare to flood irrigation. More land can be irrigated with the water thus saved.
- Crop grows consistently, healthier and matures fast.
- Early maturity results in higher and faster returns on investment.
- Fertilizer use efficiency increases by 30%.
- Cost of fertilizers, inter-culturing and labor use gets reduced.
- Fertilizer and Chemical Treatment can be given through Micro Irrigation System itself.
- Undulating terrains, Saline, Water logged, Sandy & Hilly lands can also be brought under productive cultivation.

Water Conservation Through Drip

Water is conserved in the following ways:

- Drip irrigation application uniformity is very high, usually over 90%.
- Unlike sprinklers, drip irrigation applies water directly to the soil, eliminating water loss from wind.
- Application rates are low so water may be spoon fed to the crop or plant root zone in the exact amounts required (even on a daily or hourly basis). In contrast, other methods entail higher water application quantities and less frequency. If young plants need water frequently, much of the water applied is often wasted to deep percolation or runoff.
- Low application rates are less likely to run off from heavier soils or sloping terrain.
- Drip irrigation does not water non-targeted areas such as furrows and roads in agriculture, between beds, blocks or benches in greenhouses, or hardscape, buildings or roads in landscape.

- Drip irrigation easily adapts to odd-shaped planting areas which are difficult to address with sprinklers or gravity irrigation.
- Drip irrigation is capable of germinating seeds and setting transplants which eliminates the need for “sprinklering up” and eliminates the resulting waste in the early stages of crop growth.

Drip irrigation is today’s need because Water - nature’s gift to mankind is not unlimited and free forever. World water resources are fast diminishing.

A Case Study of the Earlier Approaches to Drip Irrigation

Dating back 200 years, tribes have used bamboo drip irrigation as a means of bringing water to seasonal crops. This timeless and traditional technology uses locally available material while harnessing the forces of gravity. An assortment of holed bamboo shoots zig-zag downhill, diverting the natural flow of streams and springs across terraced cropland. The advantages of using bamboo are two-fold: it prevents leakage, increasing crop yield with less water, and makes use of natural, local, and inexpensive material.

The Jaintia, Khasi, and Garo hills of Meghalaya are largely made up of steep slopes and generally rocky terrain where the soil has low water retention capacity and where the use of groundwater channels is impossible. During the dry seasons, rainfed crops such as paddy, betal leaf, and black peppers can be irrigated by bamboo drip irrigation. Within the Jainta hills, the small village of Nongbah relies on terrace agriculture for paddy cultivation. There are no restrictions for individuals tapping into water flows from perennial streams, natural springs, or collection ponds. This enables farmers, nearly 97% of the population, to cultivate paddy, betal leaf, and black peppers in seasonal rotations. Meanwhile, drinking needs are met by perennial springs during the dry months, from October to March.

Only during the winter is irrigation required, and the bamboo system is used for crops that need relatively less water.

The few materials needed are a small dao (a type of local axe), bamboo strands of various sizes, forked branches, smaller bamboo shoots used for the channel diversions, and two willing laborers. A. Singh, in his book Bamboo Drip Irrigation Systems, investigated its use in Meghalaya and says that two workers can construct a system covering one hectare of land in 15 days. About four or five stages of irrigation zig-zag from the water source to the last point of application. Along the way, 18-20 liters of water will eventually disseminate at a rate of 20-80 drops per minute.

To get started, locate an available water source. Next, select a sloped area of land (at least 30 meters in variation). Then slice the bamboo shoots and forked branches, placing the wider shoots in the first channel and the smaller pipes for the last section (plan for 5 stages). Puncture a series of holes in the shoots, spacing them equally. Ground clearance should progressively descend so that the water may be dropped near the roots of plants in the last section (10-15 cm above ground). To reinforce the structure, tie the pipes and forked branches together using fiber-rich twine as rope. At points of diversion, smaller bamboo shoots may be used to redirect water.

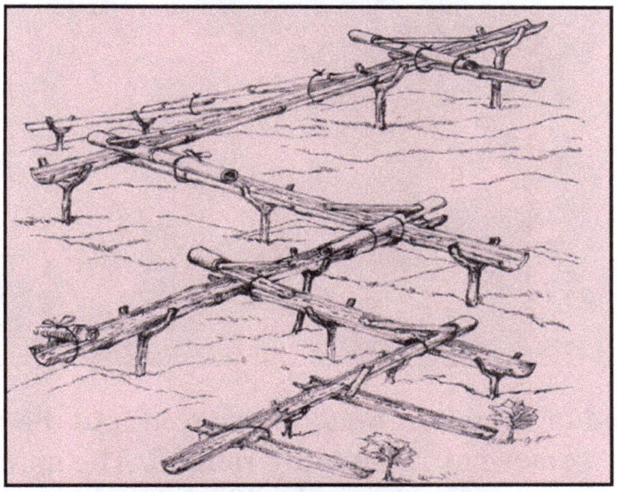

Bamboo drip irrigation system.

Materials used during installation last around three years, while maintenance is limited to cleaning and reinforcement after seasonal monsoons. Cost is also limited to labor, which can be carried out by farmers themselves. This traditional water structure is indeed contextual to location. Such variables include: replenishable supplies of bamboo, upland sources of water, and the presence of traditional terrace agriculture.

Adapting to drier growing seasons, farmers are advised to match irrigation decisions with crop selection. It is suggested that during the June to September season, rice cultivation can be stopped and farmers can opt for crops such as pulses, sunflower, sesame and maize. In areas affected severely by drought, farmers may go for pearl millet, minor millet and forage crops. However, in places where there is water stagnation, they can continue to grow rice.

In the Meghalaya hills there are an estimated 3,108 square kilometers of bamboo forests. In 1990, it was estimated that the total yield of bamboo in the state was 2.09 tons/hectare/year. The 38 different species of bamboo in the region are typically harvested at the community level and loosely managed by the Autonomous District Councils (ADC). Bamboo supplies have recently come under threat from a surge in rodent populations, gregarious flowering, disease, and large-scale extraction. Nonetheless, large-scale conservation and protection plans are underway in Arunachal Pradesh, Assam, Manipur, Meghalaya, Nagaland, Mizoram, Tripura, and Sikkim, which together contain over 50% of India's bamboo supply (within 226,000 square kilometers of land.)

The Jaintia, Khasi, and Garo hill tribes have long entrusted the use of bamboo drip irrigation as a means to fulfilling domestic, agricultural, and customary needs. Its function remains unspoiled so as the rains continue to fall and the bamboo continues to grow.

Sprinkler Irrigation

In the sprinkler method of irrigation, water is sprayed into the air and allowed to fall on the ground surface somewhat resembling rainfall. The spray is developed by the flow of water under pressure through small orifices or nozzles. The pressure is usually obtained by pumping. With careful selection of nozzle sizes, operating pressure and sprinkler spacing the amount of irrigation water required to refill the crop root zone can be applied nearly uniform at the rate to suit the infiltration rate of soil.

Crop Response to Sprinkler

The trials conducted in different parts of the country revealed water saving due to sprinkler system varies from 16 to 70 % over the traditional method with yield increase from 3 to 57 % in different crops and agro climatic conditions.

Response of Different Crops to Sprinkler Irrigation

Crops	Water Saving, %	Yield increase, %
Bajra	56	19
Barley	56	16
Bhindi	28	23
Cabbage	40	3
Cauliflower	35	12
Chillies	33	24
Cotton	36	50
Cowpea	19	3
Fenugreek	29	35
Garlic	28	6
Gram	69	57
Groundnut	20	40

Jowar	55	34
Lucerne	16	27
Maize	41	36
Onion	33	23
Potato	46	4
Sunflower	33	20
Wheat	35	24

Suitable Slopes

Sprinkler irrigation is adaptable to any farmable slope, whether uniform or undulating. The lateral pipes supplying water to the sprinklers should always be laid out along the land contour whenever possible. This will minimize the pressure changes at the sprinklers and provide a uniform irrigation.

Suitable Soils

Sprinklers are best suited to sandy soils with high infiltration rates although they are adaptable to most soils. The average application rate from the sprinklers (in mm/hour) is always chosen to be less than the basic infiltration rate of the soil so that surface ponding and runoff can be avoided.

Sprinklers are not suitable for soils which easily form a crust. If sprinkler irrigation is the only method available, then light fine sprays should be used. The larger sprinklers producing larger water droplets are to be avoided.

Suitable Irrigation Water

A good clean supply of water, free of suspended sediments, is required to avoid problems of sprinkler nozzle blockage and spoiling the crop by coating it with sediment.

Sprinkler System Layout

A typical sprinkler irrigation system consists of the following components:

- Pump unit
- Mainline and sometimes submainlines
- Laterals
- Sprinklers

The pump unit is usually a centrifugal pump, which takes water from the source and provides adequate pressure for delivery into the pipe system.

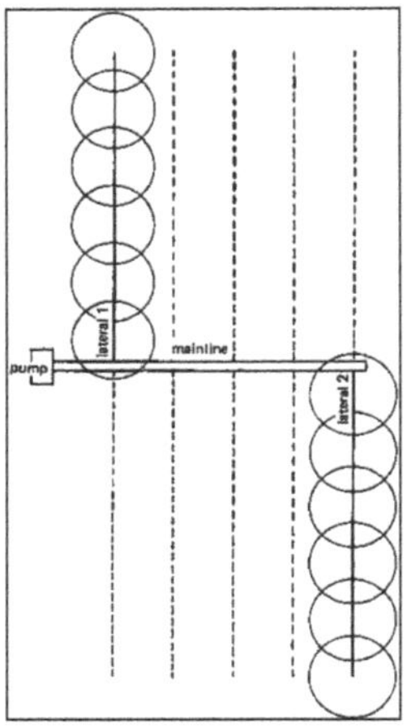

Hand-moved sprinkler system using two laterals (Laterals 1 and 2 in position 1).

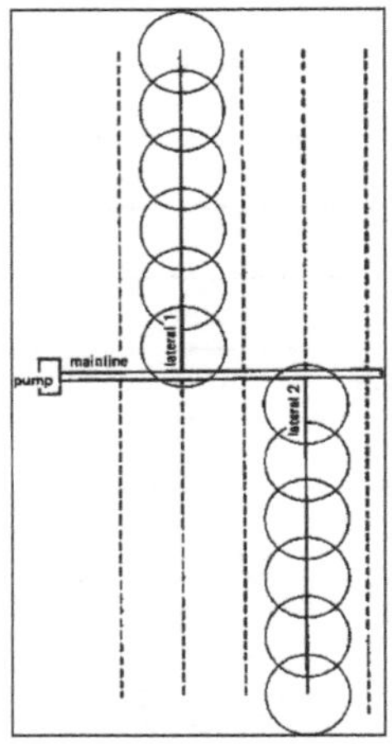

Hand-moved sprinkler system using two laterals (Laterals 1 and 2 in position 2).

The mainline and submainlines are pipes which deliver water from the pump to the laterals. In some cases these pipelines are permanent and are laid on the soil surface or buried below ground. In other cases they are temporary, and can be moved from field to field. The main pipe materials used include asbestos cement, plastic or aluminium alloy.

The laterals deliver water from the mainlines or submainlines to the sprinklers. They can be permanent but more often they are portable and made of aluminium alloy or plastic so that they can be moved easily.

The most common type of sprinkler system consists of a system of lightweight aluminium or plastic pipes which are moved by hand. The rotary sprinklers are usually spaced 9-24 m apart along the lateral which is normally 5-12.5 cm in diameter. This is so it can be carried easily. The lateral pipe is located in the field until the irrigation is complete. The pump is then switched off and the lateral is disconnected from the mainline and moved to the next location. It is re-assembled and connected to the mainline and the irrigation begins again. The lateral can be moved one to four times a day. It is gradually moved around the field until the whole field is irrigated. This is the simplest of all systems. Some use more than one lateral to irrigate larger areas.

Moving a Lateral

A common problem with sprinkler irrigation is the large labor force needed to move the pipes and sprinklers around the field. In some places such labor may not be available and may also be costly. To overcome this problem many mobile systems have been developed such as the hose reel raingun and the center pivot.

Another system which does not need a large labour force is the drag-hose sprinkler system. Main and laterals are buried PVC pipes: one lateral covers three positions. Sprinklers on risers carried by skids are attached to the laterals through hoses (similar to garden sprinklers). Only the skid with the sprinkler has to be moved from one position to another, which is an easy task.

Operating Sprinkler Systems

The main objective of a sprinkler system is to apply water as uniformly as possible to fill the root zone of the crop with water.

Wetting Patterns

The wetting pattern from a single rotary sprinkler is not very uniform. Normally the area wetted is circular. The heaviest wetting is close to the sprinkler. For good uniformity several sprinklers must be operated close together so that their patterns overlap. For good uniformity the overlap should be at least 65% of the wetted diameter. This determines the maximum spacing between sprinklers.

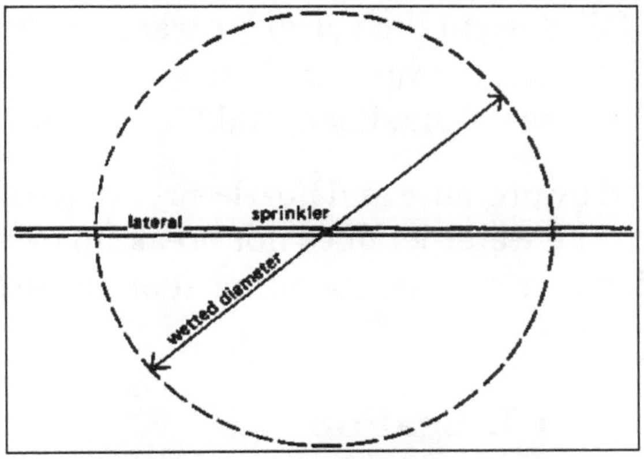

Wetting pattern for a single sprinkler.

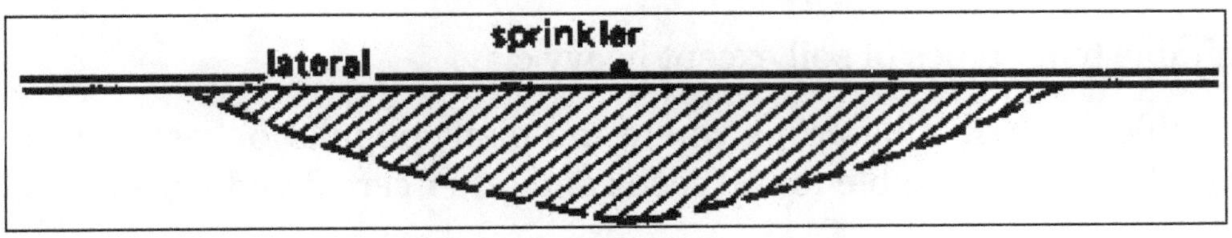

Wetting pattern for a single sprinkler.

The uniformity of sprinkler applications can be affected by wind and water pressure.

Spray from sprinklers is easily blown about by even a gentle breeze and this can seriously reduce uniformity. To reduce the effects of wind the sprinklers can be positioned more closely together.

Sprinklers will only work well at the right operating pressure recommended by the manufacturer. If the pressure is above or below this then the distribution will be affected. The most common problem is when the pressure is too low. This happens when pumps and pipes wear. Friction increases and so pressure at the sprinkler reduces. The result is that the water jet does not break up and all the water tends to fall in one area towards the outside of the wetted circle. If the pressure is too high then the distribution will also be poor. A fine spray develops which falls close to the sprinkler.

Application Rate

This is the average rate at which water is sprayed onto the crops and is measured in mm/hour. The application rate depends on the size of sprinkler nozzles, the operating pressure and the distance between sprinklers. When selecting a sprinkler system it is important to make sure that the average application rate is less than the basic infiltration rate of the soil. In this way all the water applied will be readily absorbed by the soil and there should be no runoff.

Sprinkler Drop Sizes

As water sprays from a sprinkler it breaks up into small drops between 0.5 and 4.0 mm in size. The small drops fall close to the sprinkler whereas the larger ones fall close to the edge of the wetted circle. Large drops can damage delicate crops and soils and so in such conditions it is best to use the smaller sprinklers.

Drop size is also controlled by pressure and nozzle size. When the pressure is low, drops tend to be much larger as the water jet does not break up easily. So to avoid crop and soil damage use small diameter nozzles operating at or above the normal recommended operating pressure.

Advantages of Sprinkler Irrigation

- Elimination of the channels for conveyance, therefore no conveyance loss.
- Suitable to all types of soil except heavy clay.
- Suitable for irrigating crops where the plant population per unit area is very high. It is most suitable for oil seeds and other cereal and vegetable crops.
- Water saving.
- Closer control of water application convenient for giving light and frequent irrigation and higher water application efficiency.
- Increase in yield.
- Mobility of system.
- May also be used for undulating area.
- Saves land as no bunds etc. are required.
- Influences greater conducive microclimate.

- Areas located at a higher elevation than the source can be irrigated.
- Possibility of using soluble fertilizers and chemicals.
- Less problem of clogging of sprinkler nozzles due to sediment laden water.

Subsurface Textile Irrigation

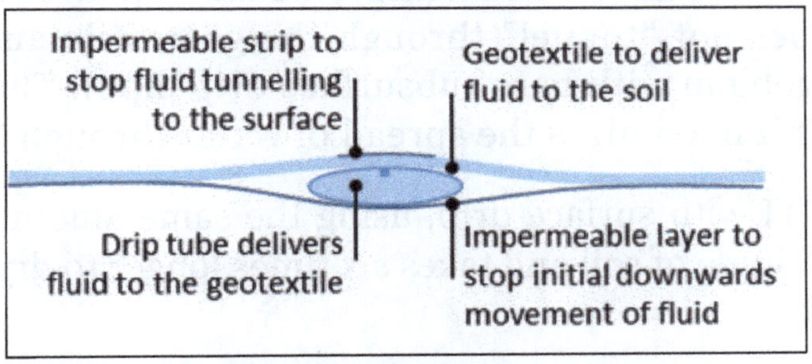

The structure of an example SSTI installation.

Subsurface Textile Irrigation (SSTI) is a technology designed specifically for subsurface irrigation in all soil textures from desert sands to heavy clays. Use of SSTI will significantly reduce the usage of water, fertilizer and herbicide. It will lower on-going operational costs and, if maintained properly, will last for decades. By delivering water and nutrients directly to the root zone, plants are healthier and have a far greater yield.

It is the only irrigation system that can safely use recycled water or treated water without expensive "polishing" treatment because water never reaches the surface.

A typical subsurface textile irrigation system has an impermeable base layer (usually polyethylene or polypropylene), a drip line running along that base, a layer of geotextile on top of the drip line and, finally, a narrow impermeable layer on top of the geotextile.

Unlike standard drip irrigation, the spacing of emitters in the drip pipe is not critical as the geotextile moves the water along the fabric up to 2m from the dripper.

SSTI is installed 15–20 cm below the surface for residential/commercial applications and 30–50 cm for agricultural applications.

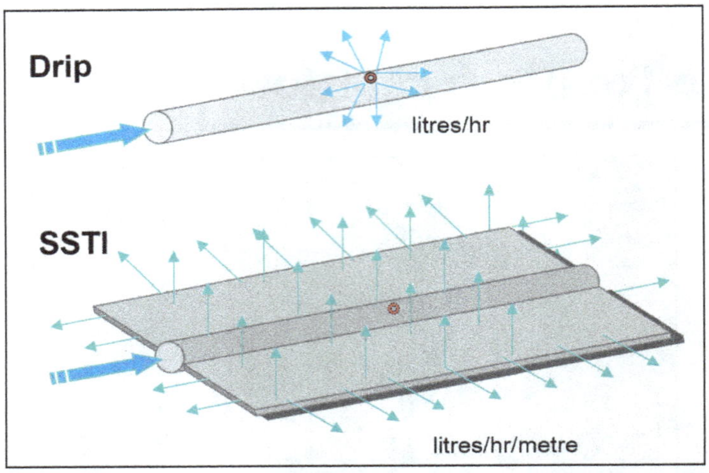

The flow of water through an SSTI installation as compared to a drip irrigation system.

To increase effectiveness, SSTI products should have an impermeable base layer to slow gravitational loss of water and to create an elliptical wetting pattern under the soil surface. It should also have a small impermeable top layer to ensure that water from the dripper does not "tunnel" through the geotextile and up to the surface (again a common problem with bare subsurface drip pipe). The effect of these two layers is dramatic as it maximizes the spread of water through the geotextile.

When comparing SSTI with surface drip, using the same amount of water, SSTI can cover 2.5 times the volume of soil and takes six times longer to dry down until the next irrigation is required.

Recycled Water and Treated Effluent

Recycled water can be used in SSTI systems as it will spread the nutrient load over 2-3 times the soil volume (compared to other irrigation methods). This means that additional nutrient requirements are minimized and the soil will have a long life without overloading other nutrients (especially phosphorus and potassium).

A major benefit of SSTI is that treated effluent can be used but is prevented from reaching the surface. Recreational or agricultural activities can continue on the field during irrigation without the contaminants coming in contact with the public.

Nutrients can be injected through all SSTI systems (fertigation). Macro and micro nutrients can be delivered to specific crops including grass, pasture, trees and vines. The nutrient is placed directly in the root zone so there is almost no wastage and no potential for run off into waterways.

Use of Geotextiles

Geotextiles were used for capillary irrigation not long after World War II. However, it was not until 1995 that extensive research was conducted at the CSIRO Division of Land and Water in Griffith, Australia, by Grain Security Foundation Ltd, that SSTI was established as a serious commercial alternative to drip.

Polyester geotextile of specific thickness and manufacture is required to ensure the system has the appropriate flow characteristics and so that it does not become hydrophobic (repels water).

Solving Typical Subsurface Problems

Studies were done on many forms of geotextile using various dripper rates and configurations to evaluate water flow for the major soil texture types and to determine if SSTI had the same problems experienced by subsurface drip systems (SDI). Specifically, considerable focus was placed on the problem of root intrusion, tunneling, dripper blockage and insect damage.

Root Intrusion

Research data showed that roots could invade the geotextile but did not thrive nor thicken in the polyester geotextile and, therefore, caused no damage. The roots were deterred from entering the dripper area because that area is dry relative to the rest of the system. Roots simply went elsewhere.

Tunneling

Tunneling (the process whereby water works its way up to the surface) was nearly eliminated by the narrow reflective impermeable tape above the drip line.

Blocked Drip Emitters

Drippers are often blocked in bare drip systems due to soil lodging in the emitters and/or soil crusting around the emitters when the system is not operating. SSTI eliminates this problem because there is a physical barrier presented by the geotextile and due to the fact that the soil remains moist for much longer than drip systems (i.e. the soil does not form a crust anyway).

Insects

Similarly, insects are deterred from damaging SSTI systems because of the geotextile barrier.

Water Delivery

Initially solid, thick-walled drip pipes were used in trials at the CSIRO but subsequent trials between 1995 and 1998 demonstrated that drip tapes (thin-walled, flexible pipe) could be used very effectively. The loss of strength from using thin-walled drip tapes was countered by the additional tensile strength of the entire system (base layer, drip tape and geotextile). While significantly reducing the cost of SSTI, the use of drip tapes also permitted the use of large diameter drip lines (of up to 35mm) allowing for run lengths up to one kilometre. The thin, flexible wall of drip tape also meant that master rolls of manufactured product could be much larger (over 600m in length).

Optimal Width and Wetting Patterns

The width of SSTI products varies based on the application. However, at the University of Queensland, Australia trials were conducted to discover the optimal width of the SSTI in black cracking clay with alfalfa sown across a single SSTI line. The SSTI line was 20 cm wide and was installed 30 cm below the surface. The alfalfa germinated directly above the SSTI line but covered 75 cm either side (a total of 1.5 m) in uniform lines. As a result of this research, most SSTI systems are between 6 cm and 20 cm wide. However, some products are up to 2 m wide.

Installation Technologies

In 1997 the first SSTI plow was tested at CSIRO Griffith using a standard three point linkage behind a tractor. This demonstrated that SSTI could be installed at 20–40 cm deep quickly and easily using the plow. SSTI plows are now available to install 1, 2 or 3 lines (more lines are possible).

Other uses of SSTI

In 1996 the first ebb and flow mat technology for potted plants was commercialized based on SSTI technology using drip tapes to control the water delivery. This ebb and flow mat form of SSTI proved very effective in producing potted plants, sprouted wheat and barley for animal production and for research purposes in producing seed varieties without the use of any overhead irrigation. It also provided an effective fertilizer delivery platform.

Components

Well-designed SSTI systems are laid out essentially the same as drip systems but in many cases SSTI can be laid in a “serpentine” pattern vastly reducing the number of take-off connections and potential leaks.

Most drip tubes/tapes used in SSTI are pressure compensating. For example, using a 16mm drip tape, a run of up to 180 m can be achieved from one connection. Longer runs of up to 1,000 m can be achieved using lower flow rates per lineal meter and/or larger diameter drip tape.

The following components make up a typical SSTI installation:

- Pump or pressurized water source to 100-300kPa (14-43psi).
- Water filters or filtration system from 120 micron with suspended solids less than 30ppm
- Fertigation injector systems.
- Back flow prevention.
- Pressure regulating valves.
- Main line. Can be LDPE or PVC.
- Solenoid valves or gate valves to control water flow.
- SSTI system laterals.
- Barbed or Spinlock fittings with stainless steel clips.
- Flushing valves at the end of laterals or combined laterals into a flushing line so that regular flushing can remove suspended solids or bacteria that may build up when using recycled water.

Using drip tapes in SSTI means that there is a wide range of commercial fittings making SSTI very easy to install. Fittings are usually spinlock or ring lock devices securing the drip tape using a barb.

Advantages of SSTI

The advantages of SSTI are:

- SSTI is a "permanent" solution if maintained properly. The components are inert and, given that they are situated below the ground, are not subject to the effects of weather, animals, machinery, vandals or other terrestrial conditions.
- Water savings of 50-75% compared with overhead systems.
- Low pressure requirement (also means lower power requirements).
- Yields can be improved up to four times in certain crops.
- Minimal root intrusion to drippers in the SSTI with a deflective tape on top.
- No emitter blockage due to crusting.
- Minimal effect of evaporation.
- Safe use of recycled or treated water.
- Can use the field (for recreation or agriculture) while irrigation is running.

- Weed growth is minimized because the water does not reach the surface (saving herbicide cost). Germination of weeds only occurs during rainfall.
- Fertigation can be done directly to the root zone (saving fertilizer cost).
- Efficient distribution of nutrients to the entire root zone.
- Broad wetting patterns (moisture covers the entire root zone).
- Water delivery can match the natural capillary rates in soil so saturation is minimized.
- Soil moisture can maintained at field capacity (minimised gravitational losses).
- No surface run off.
- Soil erosion is minimized.
- Fields do not have to be perfectly level.
- Fields with irregular shapes can be accommodated.
- Distance between lines of SSTI is far greater than drip (lower number of solenoids and other components compared to sprinklers and drip).
- SSTI laterals can be ploughed in behind a tractor on large sites (over 10,000 m per day).
- Foliage remains dry (fungal and bacterial leaf disease is minimized).

Disadvantages of SSTI

The disadvantages of SSTI are:

- Initial capital cost is typically more than overhead irrigation.
- Quality installation is critical. If mistakes made, they are difficult to find.
- Installation using correct fittings must be done.
- Regular maintenance is required to ensure long life.
- Automated control and monitoring systems are preferable (subsurface irrigation does not give any visual indicators to show if it is working or not).
- SSTI is usually not UV-treated so it must be kept out of sunlight until it is installed under the surface.
- Temporary overhead watering may be required to establish turf in hot areas.
- Germination of some agricultural crops may require overhead watering if insufficient rainfall.

- SSTI cannot apply fertilizers or herbicides overhead on the surface.
- Rodents may damage the system (although less than drip systems).

Center Pivot Irrigation

As the name suggests, center pivots irrigate in a circular pattern around a central pivot point. Pivots are capable of applying water, fertilizer, chemicals, and herbicides. This versatility can improve the efficiency of irrigation practices by using a single piece of machinery to perform several functions.

A center pivot or lateral move basically consists of pipeline (lateral) mounted on motorized structures (towers) with wheels for locomotion. A center pivot machine rotates around a " pivot" point in the center of the field whereas a lateral move machine travels along a straight path and has a separate guidance system. Sprinkler outlets are installed on the top a pipe supported by steel trusses between adjacent tower structures. The towers are usually 90 to 200 ft (30 to 60 m) apart and each tower has a 1 hp motor and sits on two large rubber or steel tires. The combination of pipe, truss and sprinklers between two towers is called a span. Flexible couplers at each tower connect the pipes of two adjacent spans. The maximum length of span is a function of pipe size, pipe thickness (strength), field slope and topography. Span length does not have to be uniform; in fact, it is often varied to match field dimensions or to provide adequate clearance between the truss and soil surface on rolling terrain. An "overhang" is a smaller pipe with sprinklers that is often suspended by cables beyond the outermost tower(s) to increase the w etted area. Large volume end guns and corner systems may also be added to the end of a machine to increase the wetted area in the corners or to cover additional areas. Machines can be more than 4900 feet (1000 m) long although the most common length of a basic machine is about 1300 ft (~400 m). The distance between the trusses and the ground can range from 4 to 14 ft (1.2 to 4.3 m) with most between 8 and 9 ft (2.5 and 2.8 m). Basic system life should be 15 to 20 years not including sprinkler heads, pumps and other ancillary equipment.

Most machines are powered by electricity, although some manufacturers use hydraulic motors, which are more expensive. A one constant speed horsepower electric or hydraulic drive motor is used to propel each tower. Tower motors should always be covered to extend their useful life. Electric power wires and/or hydraulic lines run the length of the machine with control boxes or valves at each tower. A primary control panel is usually located at the pivot base or at the engine on lateral moves. Hydraulic powered systems have a higher initial cost but may have lower annual costs because of lower maintenance and operational costs. Various manufacturers are looking at using more expensive variable speed electric motors to reduce start-stop effects on uniformity, especially with chemigation systems.

Not including a pivot base structure, there can be one to fifteen or more towers on each system. Towers are usually identified by number starting with the tower closest to the pivot base or the linear move engine/pump assembly. The towers should always follow the same tracks through the field. Equipment crossings and problems with traction as well as runoff from the compacted, wet wheel tracks are sometimes serious concerns.

Keeping linear move and center pivot machines in good alignment is critical to proper operation. Substantial damage can occur to the equipment and the crops if the alignment system fails. Alignment sensors are located on the pipeline at each tower causing the tower motor to start or stop. Alignment is controlled by electronic strain gauges, radio or laser controlled sensor systems. Often the first tower (closest to primary controls) will have an additional adjustable timer that will turn off the entire system if that tower does not move at least once every two to five minutes for extra protection against alignment system failures. Inadequate traction at a tower will often cause alignment problems.

Generally, the tower farthest from the pivot point controls the movement of the entire machine. Minimum rotation times (maximum speed) are commonly between 14 and 20 hours (2 to 3 m/min at the outer tower). Special high speed gear boxes can be installed on each tower to reduce rotation times to less than 12 hours (i.e., 4.3 m/min at the outer tower) which is often desirable on sandy or cracking clay soils. Timing controls at the control panel determine the relative average speed of the outer tower. The on-off cycle time of the outer control tower is usually about 1 minute (i.e., on a 50% speed setting, the outer motor is on 30 seconds in every minute.) A 100% setting causes the machine to travel at maximum speed (minimum rotation time) whereas a 50% setting results in the outer tower moving at half the maximum speed. Of course, the slower the rotation speed, the greater the amount of water applied. All the other towers try to stay in alignment with the end tower as regulated by the alignment system. However, tower movement in the interior of the system is somewhat random and start and stop times of 1 to 3 minutes may occur. Thus, because of the start-stop action the uniformity coefficients in the direction of travel are largest near the pivot and the end tower and the smallest near the center of the irrigation system.

Since all the towers on standard electric systems typically have the same motors and gearing ratios, the towers start and stop to stay in alignment. The start-stop action of the towers is not usually a problem with water application uniformity, which is averaged over several days. However, with pesticides becoming more specific, applied in small amounts and costing several dollars per gram, this jerky movement may present a concern if pesticides are to be applied through the irrigation water or if a separate system attached to the center pivot trusses. Costly variable speed electric motors on each tower may be justified for pesticide systems where uniformity is critical. Hydraulically powered tower motors can be adjusted so that there is no start-stop movement and applications are more uniform.

Flow requirements for center pivots and lateral move systems are defined in terms of total flow supplied to the system or the total system capacity, Q_T (gpm), for hydraulic requirements, and gross system capacity, Q_g (gpm/ac), for irrigation requirements and management. Q_g is defined as the total system capacity, QT, divided by the total irrigated area. Standard inside diameters of center pivot pipes are 5.31, 5.74, 6.38, 7.76, 8.37 and 9.76 inches (135 mm, 146 mm, 162 mm, 197 mm, 212.7 mm and 247.8 mm, respectively) with most (50-52 hectare)125 -130 acre systems using 6.38 inch (162 mm) steel tubing with a wall thickness of 0.11 inch (2.77 mm).

The two types of sprinklers used with center pivot and linear move systems are "impact" and " spray" heads. Impacts (traditional, older style impact drive) are generally low pressure, low angle (6° to 15°) heads mounted directly on the top of the pivot lateral pipe. Spray heads are subdivided into sprayers and "rotators." Sprays provide a mist or small jets that can also be mounted on top of the pipe but are more commonly installed at the bottom end of flexible drop tube connected to a U-shaped "gooseneck" on the lateral pipe. The height of the sprinklers may be adjusted throughout the season to maintain them above the crop canopy but this may negatively affect uniformity. The height, location, spacing, size and the discharge from each head are specified in the sprinkler "package" from the manufacturer. A standard 1300 ft (400 m) long center pivot will have 100 to 110 sprinklers. Low pressure spray type sprinkler heads mounted close to the canopy are probably the most popular to reduce wind and evaporation losses although low pressure impact heads on the pivot lateral are still used in some areas. Use of high pressure impact heads is becoming rare. Use of pressure regulators or flow control nozzles is very common with low pressure systems.

Water losses from spray heads near the top of the canopy typically range from 0-2% due to droplet evaporation, wind drift is usually less than 5%, evaporation from crop canopy ranges from 4 to 8%, and soil evaporation less than 2% whereas runoff may range from 0 to 15% or more depending on slope and soil conditions. Spray heads and impacts mounted on top of the pipe lateral may have droplet evaporation and w ind drift losses as high as 15%. Evaporation may slightly offset crop water use, but this amount is difficult to measure or calculate and is usually less than 15% of total ET. For spray irrigation on drops over a crop with a full canopy, application efficiencies of about 90 to 92% are attainable with no surface runoff whereas sprinklers on the top of the pipe may attain efficiencies from 80 -85% .

The first center pivots had sprinkler spacing's of about 32 ft (9.75 m) using impact heads. Later versions had variable spacing's with sprinklers closer together as they approached the distal end of the lateral. Most modern machines have a constant outlet spacing ranging from 3 to 9 ft (1 to 3 m) depending on the manufacturer and the type of system. Machines can be ordered for almost any outlet spacing although 6 to 8 ft (2 to 2.4 m) spacing's are typical, although LEPA machines may have outlets as close as every 2 to 3 feet (0.6 to 1 m). Sprinklers are installed in every outlet for linear move machines. How ever, on center pivots near the pivot where machine movement is slow, not every outlet has a sprinkler installed in order to reduce application depths. Generally, after the first tower all outlets have sprinklers installed. Uniform water applications depend on the careful matching of spacing, the particular sprinkler heads to be used and their height above the crop canopy.

Older machines and sprinkler packages often required pressures in excess of 70 psi (500 kPa) to operate, but modern machines are generally designed to operate at 35 psi (250 kPa) or less. These pressures are often too low for high volume-high pressure end guns and small electric booster pumps are often installed at the last tower. Discharge from end guns and corner systems should be controlled to avoid water applications to roads, streams or drainage facilities (especially when chemicals are being applied) by switches at the pivot base.

Pipe sizes range from about 4 inches (100 mm) to more than 12 inches (300 mm) in diameter. Pipe sizes on center pivots may decrease along the length of the lateral with increasing distance from the pivot. Linear move systems have uniform pipe sizes except for overhangs past the end towers. However, minimum pipe size is determined by strength rather than hydraulic considerations.

Regardless of the type of system, strict annual or weekly maintenance and lubrication schedules are required for all motors, gear boxes, alignment systems, couplers, seals and other parts. Sprinkler heads and pressure regulators should be replaced on a regular basis, usually every 3 to 4 years.

Lateral Move Systems

A lateral move machine travels in a straight line to irrigate up to 95% of square or rectangular shaped fields, and is supplied water through a carefully constructed open ditch that runs parallel to the direction of travel or large flexible hose-buried pipeline system. Some land is lost to production because of area needed for the supply ditch or supply hose drag lanes (30-40 ft [10-15 m] wide for the length of the field). Generally, linear move systems are used on land with slopes less than 6%.

Many lateral move systems have large diesel engines connected to a generator that powers the pump (open ditch supply) and tower motors. A hose drag system is pressurized by pumps off the field. The engine/pump/control assembly can be located in the center of the lateral move or at the edge of the field. These systems are guided by buried electrical cables, lasers or wire on stakes. All the sprinklers on a lateral move are usually the same size except for large end guns on some systems. The design of a lateral move system is a simplified case of center pivots although the management and operation is more complex with higher labor requirements. The average intensity of application, I (in/hr), for lateral move systems is given by:

$$I = \frac{K * Q_r}{W * L}$$

Where Q_T is the total water delivered to the system also called the total system capacity (gpm) , K is a units conversion factor equal to 96.25 in English units (K is 6.0 for metric units using l s-1 and m with I in mm/min), W is the width of the sprinkler pattern in feet, and L is the total length of the lateral pipe including overhangs in feet. The actual depth applied during an irrigation event for a lateral move system is given by:

$$d = \frac{K * Q_r}{L * S}$$

Where d is the applied depth in inches, K is a units conversion equal to 1.604 in English units (equals 1 .0 in metric with mm, m and m s^{-1}) and S is the speed of movement of the machine in ft/min.

Lateral move systems are often used where productive land is limited and valuable. Capital costs of lateral move machines will vary from US$500 to over US$1200 per acre (US$1250 to over US$3000 per hectare) not including water source or land costs. Labor can be a significant annual cost factor.

Center Pivot Systems

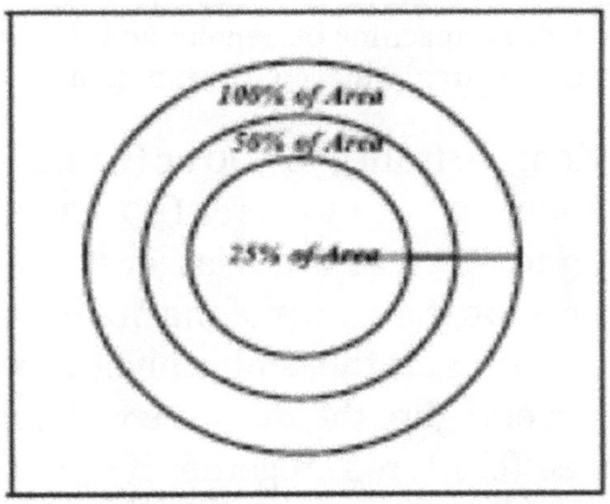

Percent of areas covered at various radii of a center pivot irrigation system without end guns or corner systems.

A center pivot machine rotates in a circle around a base pipe structure in the center of the field so that the irrigated section is any circular shape including parts of circles less than 360°. They can cover 80% to 90% of the area of a square field. Center pivots can operate on widely variable terrain with slopes as much as 30% with proper design although an upper limit of 15% slopes is generally recommended. A service road is usually necessary for control adjustment and maintenance as well as refilling, operation and monitoring of any chemigation supply tanks and injection pumps located at the pivot.

The pivot base structure is also the source of water, power and control wires. Pivot bases are usually permanent for large systems but may be portable for towable systems. Electrical power is supplied to tower motors, hydraulic and booster pumps through the slip ring connection at the pivot base.

The percent of area irrigated at various radii of a center pivot are illustrated in figure above where the innermost circle is at 50% of the radius but only contains 25% of the total irrigated area. It is important to realize that 75% of the total cropped area occurs in the outer half of the radius. Thus, management concerns tend to focus on outer towers, but many of the disease and water distribution problems will occur in the inner portions of the circle. The current state of the technology basically treats the entire field as a uniform soil and crop system. Some of the new control panels do allow changes in speed in selected sectors, but field variations are seldom pie shaped.

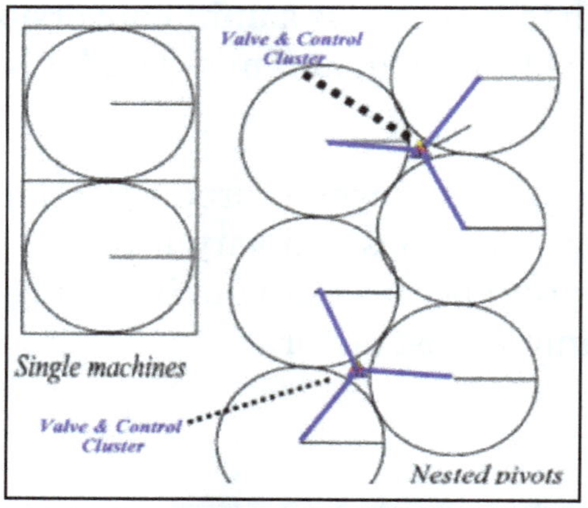

Examples if single machine placement and the increase in cropped area achieved by nesting pivots.

Center pivots are available to irrigate from 5 to 500 ac (2 to 200 ha) although a typical machine will generally irrigate about 125 to 130 acres (50-52 ha). Economic considerations usually limits their application to irrigated areas larger than 50 ac (20 ha). The area irrigated with a center pivot depends on the radius of the main lateral plus the radius increase due to end guns and corner system s. If the center pivot is positioned in the middle of a square piece of land without an end gun, the machine will irrigate about 80% of the total area. Machines are often nested (clustered) together if several center pivots are installed on one large piece of land so that 85% to 95% of the total area is irrigated.

The average operating pressure of a center pivot lateral will vary significantly depending on whether the pipeline is going up hill or down hill. This can result in large variations in sprinkler discharge so that pressure regulators or flow control nozzles are often required on every sprinkler head.

The area irrigated by a machine can be extended by the addition of relatively inexpensive high volume end guns and expensive "corner systems." Center pivot capital costs can range from US$400 to more than US$1000 per acre (US$1000-US$2500 per hectare) not including land and water development cost depending on options such as size, sprinkler packages, corner systems and end guns.

Corner Systems: Corner systems (also called corner "catchers" or swing spans) may be installed on center pivot systems to increase the irrigated, productive areas in the corners and other non-symmetrical regions along the field boundaries by adding 17-25 acres (7 to 10 ha) without buying or renting more land in areas where circles cannot be clustered. They usually consist of an additional tower and pipe system that is connected to the last tower of the main system. Corner systems normally follow behind the main system which tends to fix system rotation direction (rotation direction can be changed with great caution). The corner system tower generally has a guidance system such as phased-lock-loop circuitry that detects a low frequency radio signal from a guidance wire that is buried directly below where the tower will run. Signals are received by a high resolution antenna and receiver and fed into a microprocessor which continually monitors tracking and activates the steering motors. Sprinklers on the corner system are likewise controlled by the same microprocessor which also activates individual solenoid valves depending on their location with respect to the main end tower and the edge of the field. Use of corner systems on slopes greater than 15% may be problematic. Some corner systems use variable speed motors to improve water distribution and operation.

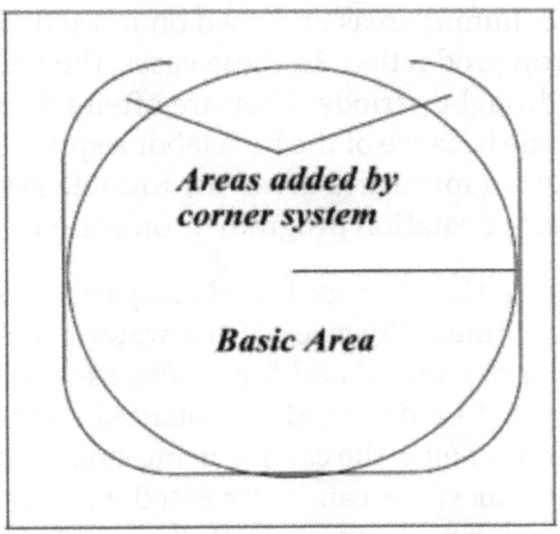

Schematic showing additional land covered by a corner system compared to a standard machine.

The incremental cost per hectare for the additional land covered by corner systems may be two to three times as expensive as land covered by the main system. Therefore, corner systems are usually only added when land values are high. In addition, a corner system can cause large fluctuations in demand from the water supply system as it turns sprinklers on and off resulting in large, undesirable pressure fluctuations and poor water distributions. Pressure and flow problems may require the installation special controls and valves or expensive variable speed pump motors. These corner system sections also tend have fairly high maintenance costs due to the complexity and many moving parts. In general, corner systems should be added only after careful analysis of all the economic benefits.

End Guns: End gun systems composed of one or more large, high pressure heads are often used to extend the area in the corners of single machines on square blocks of land. These are used as a low cost alternative to corner systems to expand irrigated area as much as 21% in corners over a 35° to 42° arc. Considering the amount of land added by a relatively small increase in radius, end guns are popular as an inexpensive way to add significant irrigated acreage to the field, however, they are not without problems. A major consideration with end guns is that they are a basically a single large sprinkler and the application depth tapers off with distance and severe drought stress may occur at the field edges. This may not be significant for biomass production (e.g., forage crops) but can be a major problem when deficit soil water conditions negatively affect crop quality. For this reason, many irrigators will turn off the end guns on their center pivots when potatoes are being grown. Booster pumps are usually required to operate the end guns which can result in the same pressure and flow problems experienced with corner systems.

Towable Systems: Towable systems can irrigate from one to four adjacent fields and are often smaller systems than fixed machines. These systems are typically used as supplemental irrigation in humid areas and used on more than one field when rainfall is insufficient for good crop production. In these cases, the machine is typically moved at least daily during the drought periods. They are often not practical in arid areas on more than one field at a time because of the high labor required for the frequent moving of them from field to field. In more arid areas the towable systems are usually moved once or twice a year as part of rotation program or on leased land.

Orchards and Vineyard Irrigation: Orchards and vineyards can be irrigated with center pivot and linear move machines. However, if the water cannot be applied below the canopy there are a number of cultural and fungal disease problems that may develop. Consequently, the fields must be designed and planted so that center pivot or linear move systems apply the water below the canopy using small sprinklers or bubblers similar to the LEPA systems. Vineyards can be irrigated with standard height machines. Orchards, on the other hand, have been successfully irrigated with high clearance (i.e., 4.3+ m high) machines.

Orchards and vineyards are generally planted in circular rows for center pivots and in straight rows for linear move systems (similar to the LEPA systems below). The irrigation machines are usually special orders to fit each specific installation. Sprinkler, spray or bubbler heads are suspended between row s on long drops with heads that apply water below the tree or vine canopy almost at ground level. As a water conservation measure and to ensure good coverage (and no cover crop between rows), there are typically two 180° flat or downward spray heads directed towards the plant row are located on each side of the plants. Fertilizers, systemic pesticides and pre-emergent herbicides can be applied with the irrigation water.

LEPA Systems

A special adaptation of the technology is the Low Energy Precision Application (LEPA) method that can be installed on both center pivot and linear move systems. LEPA has "drop" tubes about every meter that extend to the soil surface where a low-pressure bubbler is attached in place of a sprinkler. Water is applied directly to the furrow and evaporation losses are minimized since the canopy is not wetted. These systems can be very efficient (e.g., 95- 98%) since evaporation losses (soil evaporation generally less than 2% with alternate row irrigation, although runoff may be as much as 50 % with poorly designed and operated systems) are minimal although initial capital costs are higher than standard systems.

Crops are usually planted in a circle so that the drops do not damage plants. Sometimes a canvas "sock" or other fabric energy dissipation device is used to prevent soil erosion in the furrows. The use of a machine such as the Dammer-Diker™ is often used to create small reservoirs to store water until it has infiltrated on heavy or steeply sloping soils under both LEPA and regular application techniques. Typical quarter mile long (400 m) LEPA systems will have 350 to 450 heads.

Operational Characteristics of Center Pivot Irrigation Systems

Many consider the current center pivot technology to be mature. They are mechanically reliable, simple to operate and require little supervision. However, the management for these systems is much different and unique compared to other types of irrigation systems. These systems are inherently characterized by light, frequent irrigations (e.g., daily) which offers numerous water and nutrient management advantages as well as numerous cultural disadvantages.

From a benefits standpoint, water and water soluble nutrients can be carefully applied in amount to exactly meet plant needs. The light applications can potentially reduce leaching in sandy soils (or cracking clays). Culturally, the frequent wetting of the canopy often creates ideal conditions for many fungal diseases, especially inside the tower closest to the pivot base. Shallow root development is encouraged on many crops by the frequent, light irrigations so there is often little buffering against drought stresses in the event of a mechanical breakdown of the system. For this reason, soil water contents in the upper regions of the root zone generally have to maintained at relatively high levels.

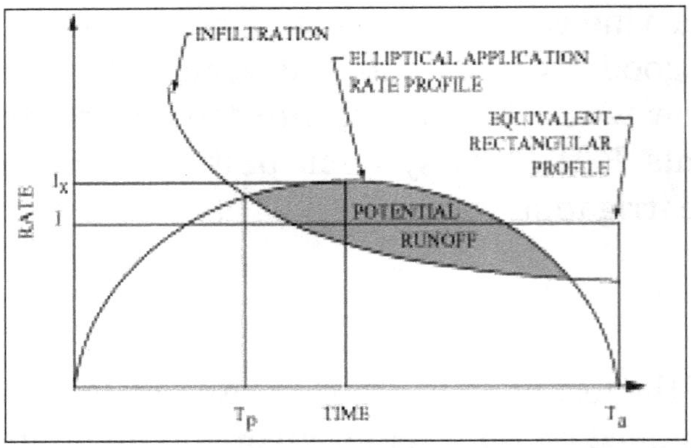

Interaction between the infiltration function and the application rate pattern (Tp is time to ponding, Ta is the application time).

The frequent irrigations require adjustment of rotation times such that the machine is not in the same spot in the field every day at the same time to "average" losses and over applications across the field over time. Thus, rotation times that are 12 hour multiples are avoided.

Matching Application and Infiltration Rates

A major physical phenomena that center pivots take advantage of is that initial water infiltration rates into soils are high. Light, quick applications take maximum advantage of this phenomenon. To illustrate, the outermost tower of a basic 125+ ac (50+ ha) center pivot can travel 3 to 15 ft per minute (1 to 4 m min-1). However, the innermost tower travels only about 10% of that speed. This means that sprinklers at the outer tower are applying water 10 times faster than those near the first tower in order to have the same depth of application applied along the entire length of the pivot. With some sprinkler packages the application rate at the outer tower may exceed 4 in./hr (100 mm/hr).

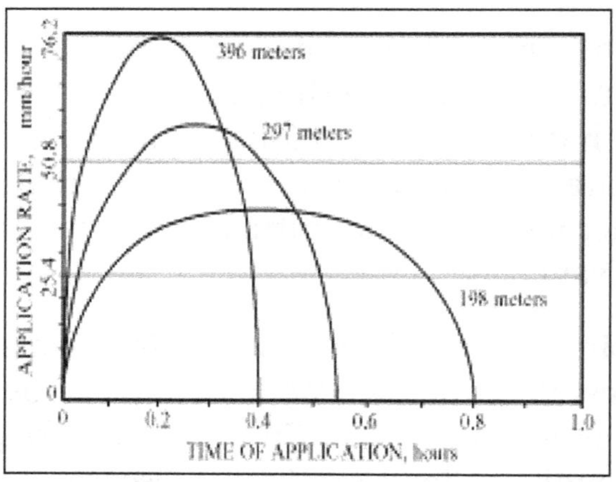

Application rate profiles and time of wetting showing the wetted diameters of the sprinkler heads depending on their position along the lateral length relative to the pivot base with a uniform width of wetted strip.

Thus, the sprinklers at the end of the machine generally cover larger diameters even at higher system rotation speeds to avoid exceeding the infiltration rate of the soils. Figure below shows the interaction between the application rate and the infiltration function. The intensity of application is illustrated in figure below relative to position in the field assuming the same amount of water is being applied showing the different wetting times. The rate of water application reaches a peak when the sprinkler passes directly over a location at a, b and the full radius of a basic system. The objective of proper nozzle selection and system operation is to ensure that the application rates do not exceed the respective infiltration rates at various points along the lateral.

In order to meet the application depth requirements, the discharge from sprinklers basically increases linearly with the radius as shown in figure above. In addition, the majority of the pressure loss occurs in the first one-third of the lateral pipe. A sprinkler at 985 feet (300 meters) from the pivot base will have twice the discharge of a sprinkler at 492 feet (150 m). Thus, the required discharge for any individual sprinkler along the pivot lateral is:

$$q_s = \frac{K\pi RSQ_g}{K}$$

where q_s is the individual sprinkler discharge in gpm, R is the radial distance from the pivot base in feet, S is the spacing between adjacent sprinklers along the pipe lateral in feet, and Qg is the gross system capacity for the irrigation system in gpm per acre, K is a units conversion of 43560.0 in English units (equals 10000. 0 in metric units of L s-1, m, and L s-1ha-1). Special soil cultural practices may have to be implemented if the system capacity results in runoff in various areas of the field. Table below presents sprinkler discharge requirements with length as a function of gross system capacity.

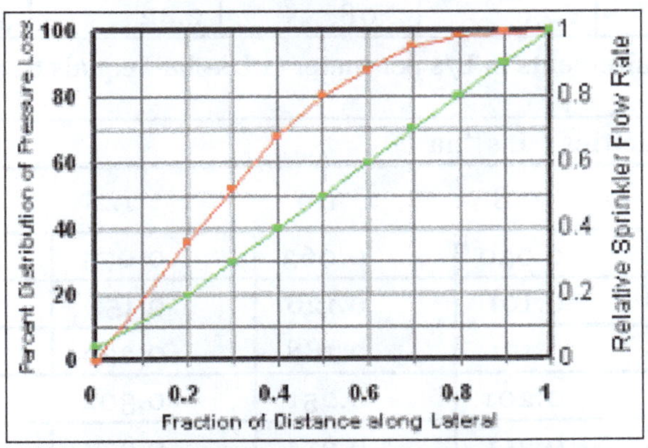

Typical distributions of pressures (upper line) and flow rates (lower line) from individual sprinklers along the lateral length.

Assuming uniform crop, soil, microclimate and topographic conditions, the goal of irrigation is to have the most uniform water application pattern possible. Two major variables with selection of sprinklers are spacing and the type or size of the heads. Sprinklers must be spaced close enough to have good overlapping of wetting patterns. The sprinkler type and discharge must be selected to avoid runoff. They must be matched to the soil and crop. The kinetic energy and power with which droplets impact the soil can have a large effect on soil compaction and sealing which can greatly increase runoff (a major problem with end guns), and small droplets may be beneficial in reducing soil sealing. However, small droplets are subject to wind drift and evaporative losses. Thus, selection of best sprinkler heads is, at best, a compromise between often-conflicting criteria and additional measures such as creating small storage reservoirs in the furrows may be required.

Generally, nozzle sizes are small near the pivot base and gradually increase in size and discharge as the radius increases. Corner systems generally have similar or slightly larger nozzles as the end tower of the basic system. On very long systems, the largest nozzles may not have enough flow capacity and two or more sprinklers must be installed or the system flow requirements reduced and system pressure increased. Infiltration problems are often reduced by spreading water applications over are larger area through placement of drop tubes on alternating sides of the pivot lateral truss structure or having two to four small heads on a small boom mounted almost perpendicular to the pivot lateral.

	System Capacities (gpm/ac)					
Radius, ft	4.0	5.0	6.0	7.0	9.0	10.0
300	0.17	0.22	0.26	0.30	0.39	0.43
600	0.35	0.43	0.52	0.61	0.78	0.87
900	0.52	0.65	0.78	0.91	1.17	1.30
1200	0.69	0.87	1.04	1.21	1.56	1.73
1500	0.87	1.08	1.30	1.51	1.95	2.16
1800	1.04	1.30	1.56	1.82	2.34	2.60
2100	1.21	1.51	1.82	2.12	2.73	3.03
2400	1.38	1.73	2.08	2.42	3.12	3.46

Sprinkler discharge requirements in L/s per meter (1 L $s^{-1}ha^{-1}$ equals 6 .414 US gallons per acre).

	System Capacities (L $s^{-1}ha^{-1}$)					
Radius, ft	0.6	0.8	1.0	1.2	1.4	1.6
100	0.038	0.050	0.063	0.075	0.088	0.101
200	0.075	0.101	0.126	0.151	0.176	0.201
300	0.113	0.151	0.188	0.226	0.264	0.302
400	0.151	0.201	0.251	0.302	0.352	0.402
500	0.188	0.251	0.314	0.377	0.440	0.503
600	0.226	0.302	0.377	0.452	0.528	0.603
700	0.264	0.352	0.440	0.528	0.616	0.704
800	0.302	0.402	0.503	0.603	0.704	0.804

Soil infiltration functions must be determined prior to the design and nozzle selection process and related to opportunity time. The opportunity time is less towards the outer towers since the sprinklers are traveling faster which must be offset by higher flow rates and wider wetted diameters per head with increasing distance from the pivot base. Thus, to have the same application depth, sprinkler selection and rotation speed are dependent on bare soil (no crop) and topographic factors to avoid soil erosion and wasting water and energy. The speed, S_j, at point j on the radius r is $2_j/ t_{rotation}$ where $t_{rotation}$ is the time required for a complete rotation in minutes. The infiltration opportunity time t_j is defined as the wetted diameter of the nozzle, W_j, divided the speed, S_j. The average intensity (application rate in./hr), I_j, of application at radius (r) at point j which can be calculated as:

$$I_j = \frac{d_g}{t_j} = \frac{K r_j\, Q_g\, R_e}{W_j\, L^2}$$

where K is a units conversion factor equal to 192 .3 for English units (7200.6 in metric units of l s^{-1} and m), Q is the gross system capacity for a basic circle (no end guns or corner systems) in gpm, L is the radius of the basic circle in feet and W_j is the width in feet of the sprinkler application pattern at j. R_e is a loss factor equal to 1.0-fraction of estimated evaporation and wind drift losses (e.g., 0.10 to 0.15). Thus, this equation must be solved in a trial-and- error procedure until the selected nozzles and minimum system rotation speed do not cause runoff. Use 50% or less of the maximum system rotational speed for these calculations to allow for changing conditions through the season and management flexibility. Thus, the average application rate, I_L, in in/hr at the end of the basic pivot, L, is:

$$I_L = K\frac{L d_g}{W t_{rotation}}$$

where $t_{rotation}$ is in hours, dg is the gross daily application depth in inches, L and W are in feet, and K is a units factor of 6.2832 (same in metric units of mm/hr and meters). The average depth of water applied, d_g, per revolution is given by:

$$d_g = K\, Q_g\, h_r$$

where K is the units conversion equal to 0.00221 (K is 0 .36 in metric units of mm and l $s^{-1}ha^{-1}$), Q_g system capacity in gpm/ac, and hr is the hours per revolution.

The speed control setting, C_s, is a ratio of the depths at 100% speed (minimum depth per revolution) to the desired depth at a lower speed. It can also be determined as the ratio of the depth per revolution at the 100 % speed, d_g, to the application depth per revolution at a given speed, S, in percent, which is:

$$C_s = K\frac{R_t * v_m * d_g}{Q_r}$$

where K is a units factor equal to 31.168 (0.84 in metric units of l s^{-1}, meters, meters per minute and mm per revolution), Q_T is the total system capacity in gpm, R_l is the distance from the pivot base to the end tower in feet, $_m$ is the velocity of the end tower in feet per minute for that revolution, and d_g is in inches applied at the 100% speed setting that should not result in runoff . Substituting equation 6 into equation 7 gives C_s in terms of velocity and hours per revolution (K would be 0 .01667 in both English and metric units).

System Selection Considerations

Because center pivot technology is so well developed, it may seem that design concerns are minimal. However, this is absolutely not the case. The machines must be designed to match each site. Information must be collected to characterize the variability of the soils (physical and chemical), topography, infiltration rates, microclimates across a field, and expected crop water use patterns over the season. Water supply (quality, quantity, timing and long term availability) must be investigated as well as any other potential physical, legal or social constraints must be identified. Steepness of slopes along wheel tracks can affect performance and long system life. Potential losses such as wind drift, evaporation, runoff, and deep percolation need to be estimated. Sprinkler spray and distribution patterns characteristics must match the soil. Machine operational criteria and strategies must be developed.

Total system flow requirement is needed for hydraulic calculations and proper selection of pumps, main lines and lateral sizes. The calculation for total system flow, Q_T (gpm), is:

$$Q_T = K * A * d_g$$

where K is a units factor of 18.86 (0.1158 in metric units of L s^{-1}, hectares and mm), A is the total area irrigated in acres and d_g is the daily gross depth applied per unit area in inches over the irrigation period of 24 hours. The total system flow, Q_T, can also be calculated based on the amount of water to be applied over a fixed time period, f, as:

$$Q_T = K \frac{A * d}{f * T}$$

where f is in days, d is the average depth applied in inches, T is hours per day of operation, and K is a units factor equal to 452.6 (use 2.78 in metric units of l s^{-1}, hectares and mm).

The gross system capacity, Q_g, is a useful quantity for easy comparisons of the adequacy of design and management of different sizes of center pivots and lateral move systems. It should be the amount of water continuously delivered to the machine per acre that is sufficient to meet peak evapotranspiration (ET) requirements as w ell as losses. Q_g typically ranges from 4 gpm/ac (0.65 l s^{-1} ha^{-1}) to as much as 10 gpm/ac (1.6 l $s^{-1}ha^{-1}$) although the average is around 6 to 8 gpm/ac (1.0 to 1.2 l $s^{-1}ha^{-1}$) which allows some flexibility in case of breakdowns. Some systems are specifically designed to operate at a deficit during the peak water user periods to stretch water supplies or add additional center pivots to increase the total irrigated area, but this should be done only after careful consideration of all factors. A properly selected Q_g should account for local crop, climate and soil characteristics.

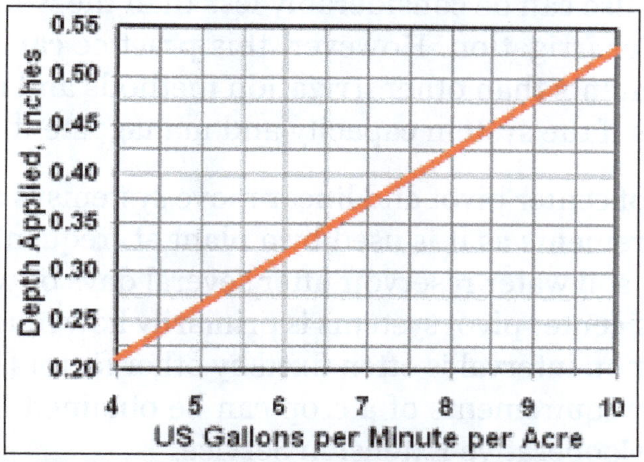

Water application depths in a 24 hour period for center pivots as related to the gross system capacity with no losses (100 % efficiency).

Some systems are deliberately under designed (gross system capacity too small) in order to stretch water supplies and irrigate more area, realizing that the plants may be stressed during peak water use requirement periods. Designing for less than 6 gpm/ac (~1 L $s^{-1}ha^{-1}$) is not recommended in arid areas of the Pacific Northwest since many crops would be stressed during peak water use periods. A design gross system capacity value of 8 gpm/ac (1.2 L $s^{-1}ha^{-1}$) is more appropriate.

Irrigation Scheduling

Center pivots and lateral move sprinkler systems will usually use more water than other sprinkler methods because the frequent irrigations have more evaporation losses from the plant canopy and soil as well as wind drift which occur with every application rather than once every 7 to 10 days. LEPA systems will be more efficient. Irrigations should be scheduled based on soil water levels to avoid undesirable levels of crop stress. This is compounded by the light frequent applications, shallow rooting and cultural operations such as fertigation, spray and tillage programs. If system capacity is not adequate to meet peak water use requirements, it may be necessary to build up soil water reserves and encourage deeper rooting prior to any water short periods, however, this it is often difficult to limit over-watering and avoid deep percolation losses.

Pressure variations due to static topographic differences and changes in system flow rate by the operation of end guns or corner systems can easily result in greater than 10 % variations (±5%) in water applications across a field. These differences in application depths are small on a daily basis, but the effects are additive over the season and may have serious consequences on water sensitive crops such as potatoes. Thus, the actual daily variation in application depths and the actual distribution of water must be considered in irrigation scheduling. General irrigation scheduling concerns are presented in the adjacent box.

Water Requirements. Center pivot and linear move systems inherently provide frequent, small water applications. Consequently, the volume of water stored in the soil and available for crop use can be considerably less than the wetted soil volume under other types of sprinkler irrigation. However, this practice can maintain higher, less variable soil water contents than other irrigation methods and reduce the occurrence of plant water stresses if the system capacity and management are appropriate.

The basic philosophy of center pivot and linear move systems is to replace water in the root zone in small increments as it is used by a plant at frequent intervals rather than refilling a much larger soil water reservoir after several days or weeks. Thus, the major concern for scheduling center pivot systems is primarily how much to apply during irrigation since the irrigation interval is often fixed by other factors, including design. The actual seasonal water requirements of a crop can be obtained from various technical sources including the Cooperative Extension Service.

The gross required depth, dg, in inches (in metric use mm per day of ET and P) of application per day is given by:

$$d_g = \frac{k_f\, ET_c - P_e}{E_a\ /100}$$

where k_f is a frequency factor to adjust the actual daily crop evapotranspiration (Et_c) in inches during the period for high frequency water applications which is typically about 1.2 for daily irrigations, 1.1 for irrigations every two days and 1.0 when irrigations are every 5 days or more. ET_c can be calculated from the Modified Penman, Penman-Monteith or other suitable estimating equation depending on data availability. This adjustment is necessary because of increased evaporation losses from the plant canopy and soil surface inherent in high frequency irrigation regimes. P_e is the effective precipitation during the period in inches and E_a is the application efficiency as a percent.

Using the same variables as previously defined except that d is the net depth applied, the maximum time between,

$$f = \frac{d}{k_f\, ET_c - P}$$

irrigations, f, can be calculated as:

The estimated crop water use (ET) combined with amount of the area to be irrigated, will determine the total volume of water to be applied in an irrigation subject to system capacity. The system should be able to apply the maximum depth of water needed during peak water use periods accounting for all losses. The maximum interval between irrigations is primarily controlled by soil hydraulic characteristics, soil profile layering, and maximum allowable deficit levels for the crop. The depth of root zone soil, saturated hydraulic conductivities and soil water holding capacities may the volume applied in a single irrigation to avoid runoff or excessive deep percolation.

It is sometimes not possible to achieve optimum irrigation schedules because of irrigation system limitations. These include inflexibility in controls and instrumentation, inadequate system hydraulic capacities (including fill times and system drainage), and the quantity and quality of available water throughout the season.

Management considerations such as the quality and quantity of available labor can affect the ability to implement irrigation schedules. Timing, amount and label requirements for chemigation may also negatively influence optimum schedules. Irrigation schedules may have to be adjusted because of cultural or harvesting considerations or to take advantage of lower "off peak" electrical power rates.

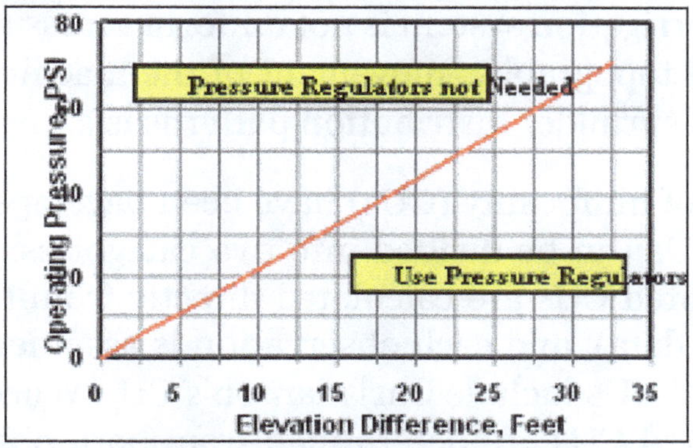

Suggested recommendations for the determination of the need for pressure regulators on center pivots and linear move systems.

Depth and Pressure Distributions

Sprinkler flow rates will vary as the lateral rotates on sloping land unless pressure regulators are used. Pressure regulators will be needed on each sprinkler head if the pressure varies by more than ±10% oven the length of the pivot at any point in the field. Pressure regulators are almost always required on low pressure systems on sloping land. Figure above shows general recommendations on whether or not pressure regulators are needed on a system. The selection of specific regulators depends on system pressures and the sprinkler selection. Flow control nozzles may also be an option but large fluctuations in pressures may adversely affect distribution patterns and droplet sizes.

Selection of the proper nozzles requires knowledge of the pressure distribution along the pivot lateral. This is complicated by the fact that the sprinkler discharges increase as you move toward the end of the pivot lateral while the pipe diameter remains constant (e.g., 6.38 inches [162 mm] diameter until the overhang after the last tower). The pressure, P_j, at point j in psi along the lateral is given by:

$$P_j = P_o - \frac{P_{lp} * L * f_p(R)}{100} - K\,E_g$$

where P_o is the pressure at the inlet to the pivot in psi, P_{lp} is the pressure loss in the pivot lateral pipe in psi per 100 ft, L is the radial length in feet to j, E_g is the elevation gain in feet at j, K is a units conversion factor of 0.4484 (use 0.1017 with meters and kPa) and f_p (R) is the dimensionless pressure distribution factor at distance R (at j). R is about 0.555 for most center pivots without an end gun and 0.56 with an end gun. A value of 0.36 is used for R on linear move systems due to the more equal distribution of flow from the outlets along the lateral. This can relationship also be calculated by the Hazen-Williams equation using a C factor of 140 or 145 for galvanized or epoxy lined steel pipe.

Distribution Patterns

Water application uniformity is an important performance criterion for the design and evaluation of center-pivot sprinkler irrigation systems. However, the water application depth of a center-pivot irrigation system is not uniform across a field. It depends on the sprinkler package, field topography, movement of the machine, and many other factors. Wind distortion of sprinkler distribution patterns is a major and dynamic factor.

Numerous coefficients of uniformity (CUs) have been developed over the past few decades. In general, all CUs can be divided into two categories: non-weighted and areal-weighted. Non-weighted CUs are calculated directly from the observations (actual or simulated catch-can data), and each observation is assumed to represent the same land area. Non-weighted CUs include Christiansen's CU_C, Wilcox and Swailes, Benami and Hore. Areal-weighted CUs are determined from both observations and the land area each observation represents. Areal-weighted CUs include CU_H by Heermann and Hein and the USDA-Soil Conservation Service pattern efficiency. For irrigation systems where each sprinkler covers the same amount of land area, non-weighted CUs can be used. On the other hand, if each sprinkler in an irrigation system covers different sized area, such as in a center-pivot irrigation system, areal-weighted CUs are preferred. The uniformity coefficient can range from 0.0 to 1.0, but the minimum desirable uniformity is about 0.85 for a center pivot irrigation systems. Both the application efficiency and uniformity coefficients are affected by the depth of irrigation.

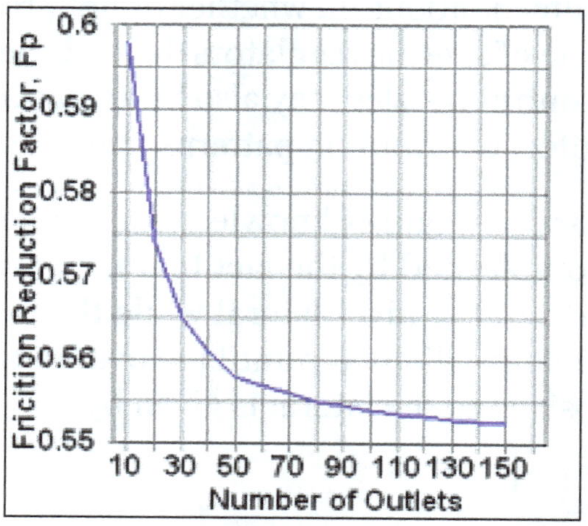

Friction reduction factors for multiple outlet pipelines.

Although a coefficient of uniformity can be used to compare different systems, it does not provide a functional description of actual applied water distribution. Therefore, statistical distribution functions are often used to represent the actual water application distribution. Many distribution functions have been proposed. Among them is the normal distribution, log normal, uniform and specialized power distributions.

One advantage of using distribution functions over a CU is that matching raw data to a theoretical function results in better estimation of the performance compared to the direct use of the raw data. When a distribution model is known, raw data are used to fit the distribution function and to obtain the function coefficients; and the coefficient of variance (CV) is calculated and used to compute the CU. The CV approach has been stated to be more appropriate than the CU approach in some cases. Distribution functions can also be used to determine the volume of over and under irrigation. This is done by simply integrating the product of the application depth and depth distribution function.

In using a distribution function to evaluate a sprinkler system design, two important points must be kept in mind. First, the distribution function is the probability distribution, not the actual spatial distribution, of water application depths. It can tell the probability of a given application depth, but it cannot tell which locations of the field actually receive the given amount of water. A distribution function may be adequate for evaluating the overall system performance, but it is not sufficient when a spatial water distribution is required as is needed, for example, in managing chemigation. Secondly, the distribution function is developed based on the assumption that water application depth is a random variable. That assumption may not be true if sprinkler distribution patterns have highly predictable shapes. In such a case, application depth may be directly calculated, and assuming a random distribution of application depth is vulnerable to errors and misinterpretations.

A center pivot irrigation model (CPIM) was developed to study the non-uniform distribution of irrigation water/nitrogen from the center pivot system. CPIM is a basic hydraulic model that considers topography and predicts nozzle pressures at any water emission point in the field. An empirical shape-pressure function specific to each sprinkler head is used to predict the water distribution from each head. CPIM overlaps the patterns and can produce maps showing the spatial distribution of water application depths. Complete field information on a center pivot system including actual topography and hydraulic data can be entered into the program's databases. There are also several "default" sprinkler (impacts only, at this time) packages and topographic options that can either be used directly or edited. CPIM is currently structured to fit within our defined GIS framework.

The application uniformity of the whole field can also be assessed. The CPIM results can be graphically compared with actual catch-can test results, if available. Results of the CPIM analysis could be used to target areas of highest potential nitrate leaching within a field for specific management practices.

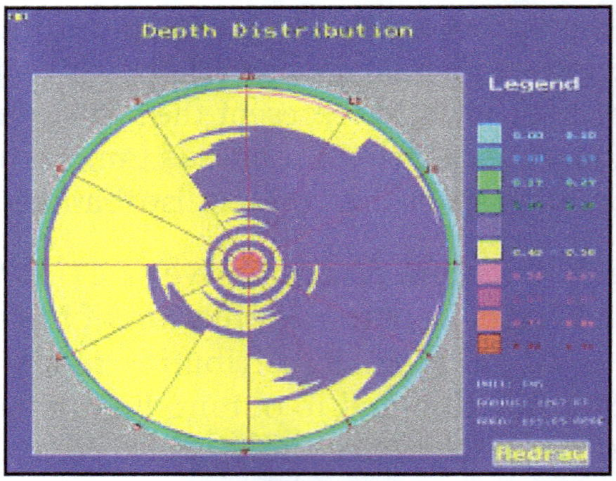

Example of water distribution from the CPIM model on steeply sloping field without pressure regulators.

Chemigation

Center pivots provide an excellent vehicle to apply some chemicals and many fertilizers to exactly match plant requirements. In some areas with very light soils as much as 80% of nitrogen fertilizer is applied through the center pivot system. Substantial crop quality and pest control benefits may accrue when using this method.

The first rule of chemigation is safety. Special chemigation safety devices, check valves and air relief valves are required for all chemical injection systems under federal and state regulations. Well heads must be protected from reverse flows, system drainage or back siphoning. Electric and hydraulic interlocks with time delays must be installed between the injectors and irrigation pumps to prevent chemical injection when the irrigation system is not operating.

Personnel must be specifically trained and, in many areas, licensed for chemical applications. Injection of any pesticide into an irrigation system must be specifically permitted by the pesticide label and may also be subjected to additional state regulations. Detailed records of all chemical applications need to be maintained for safety, evaluation, legal and regulatory requirements.

All chemicals and chemical-water mixtures must be checked to avoid phytotoxic effects before any injection occurs. In addition, it is critical that all the chemicals being injected at one time are compatible with each other and the water chemistry and concentration limits are not exceeded so that precipitates do not form. Emulsifiable pesticide concentrates and wetable powders may require special design and management considerations (e.g., mechanical supply tank agitation) to help ensure uniform applications. Acidification to lower water pH may sometimes be required prior to injection of the chemical.

Injection installations should always provide for complete mixing and uniform concentrations before the chemicals reach the field. Materials should be injected into the center of the water flow to ensure quick dilution to reduce deterioration of piping, valving or other components. Generally, injection rates should not exceed 0.1% of the system water flow rate although concentration limits and label requirements for pesticides are usually less.

The use of positive displacement pumps is highly recommended for liquid chemical injections. The pumps should be adjustable and able to inject any water soluble chemical at low concentration levels (e.g., #100 ppm). The use of an in-line mixing chamber after injection may be necessary in some cases. A flow meter or other flow detection device can be connected to a controller that is programed to inject a specified amount of chemical from a nurse tank into the irrigation system at specific times based on flow rate.

Maintenance

Because center pivot and lateral move systems inherently supply small amounts of water on a frequent basis, soil water storage is usually limited. Consequently, an extended breakdown can be very serious in terms of yield reduction and expensive service calls.

Some system maintenance will always be required during the irrigation season. This typically involves replacing or repairing sprinkler heads or sprayers, leaky valves, flat tires, electrical shorts, oil leaks in gear boxes and breakdowns in gear boxes or drive line U-joint couplers.

Many in-season problems could be averted by a strong preseason maintenance program that includes: checking flanges, rubber flex boots, collector ring base drain and system drain valves for leaks; tightening nuts and bolts on flanges, trusses, tower supports, flex boot bands, lug nuts, etc.; greasing the pivot swivel, changing oil in gear boxes and replacing leaky gaskets as necessary; cleaning the pump panels and screens; cleaning or replacing filters, screens and ventilation/drain holes on engines, gear boxes and electrical panels; checking electrical systems including power cables, grounding conductors, power and pump shutdown wires, swivel connectors, pivot contactors; testing, replacing and repairing defective sprinkler heads, end gun bearings and pressure regulators; and checking that system operational water pressures are appropriate and pressure gauges are accurate.

Aggressive nozzle "management" and leak prevention programs can save water and energy. Nozzles become worn by silt and sand particles in the irrigation water leading to higher flow rates, less efficient pump operation and possible decreases in system pressures. Replacing worn nozzles on a regular basis, checking that the appropriate nozzles are installed in the correct locations and the installation of flow control nozzles/pressure regulators where needed will help ensure good uniformities of application and reduce overall water use.

Special Hardware Selection Considerations

Control Panels and Communications

There are a tremendous number of options available for control and operation of center pivots. Panels can be operated and adjusted manually or remotely by phone or radio. Rapid advances in computer and communications technologies have made it possible to remotely monitor as well as turn machines and injection pumps on and off from the farm office or even the front seat of the grower's pickup truck. As many as 100 or more machines may be controlled at one time. Alarms may sound when a machine stops for a multitude of reasons.

All the main manufactures are supplying advanced control panels at the pivot with remote communications capabilities. These are primarily digital devices with minimal electro-mechanical parts. Graphical interfaces can set system speed, run the system wet or dry, rate of application, system direction, end gun valves, fertilizer pumps, and other ancillary equipment. Different rotation speeds can be easily pre-programmed for different sections of a circle. The control panel will also keep a digital record of events for later analysis and record keeping.

These technologies are not inexpensive and the irrigator must determine the economic tradeoffs when selecting which panel and communication system (if any) will be purchased. Growers must evaluate their operation and maintenance costs and compare these with anticipated productivity gains and/or labor savings from various remote control options.

Tire Selection and Gear Boxes

Proper tire selection is critical to avoid problems with traction, deep ruts and easy crossing of the wheel tracks by farm equipment. Tires will vary with width and diameter ranging in size from about 24 to 38 inches (610 to 965 mm) in height and 11 to 17 inches (280 to 430 mm) in width. Generally, narrow larger diameter tires are used on heavy clay or loamy soils while wider tires are used on lighter sandy soils for greater flotation. Row crop farmers will often choose narrow tires where as growers of permanent crops such as alfalfa will normally select wide tires. Turf or tractor treads can be ordered although steel tires with heavy lugs or steel traction rims with heavy duty lugs mounted inside a regular rubber tire may be used on heavy clay soils where traction is a concern.

Obtaining good flotation can be a compromise of tire width and diameter. A narrow large diameter tire may have the same soil contact area or "foot print" as a wider small tower and provide equal flotation. Narrower tire tracks are usually easier to cross with equipment.

Gear boxes selection can be difficult. Worm gear systems are more expensive used on widely variable, steep terrain situations and on heavy soils with traction problems. Planetary and spur drive gear boxes are used in less demanding situations.

Screens and Sand Traps

When water is pumped directly from rivers, lakes or canals, the intakes should be equipped with self-cleaning screens. The stainless steel screens should be about twice the diameter of the attached pipeline and a mesh opening of about 0.25 inches (6 mm) or less. Cleaning is often accomplished with internal pressurized water jets that rotate inside the intake screen and push debris away from the mesh openings.

The pivot point should also have a stainless steel or galvanized screen with a mesh size of 0.1 inch (3 mm) or less to keep debris, algae, weed seeds, etc. from plugging nozzles. There needs to be a way to hydraulically isolate the screen from the rest of the system. These screens can be self-cleaning or manually cleaned. There should be a pressurized water supply for a hose to manually wash the screen in both cases.

Since sand and small gravel tend to collect at the distal end of pipelines, sand traps should also be placed at the distal end(s) of the center pivot or linear move system. These typically consist of a short section of 4-inch (100 mm) pipe pointing downward from a tee near the end of the mainline (near the end gun). These pipes have a 4-inch (100 mm) spring loaded valve or other method to quickly flush the collected sand from the system. Sometimes a special hose and large diameter (e.g., 0.25 in [6 mm] diameter) nozzle-spray plate arrangement is used to continuously flush the sand while the system is operating.

Garden Hose

A garden hose, hosepipe, or simply hose is a flexible tube used to convey water. There are a number of common attachments available for the end of the hose, such as sprayers and sprinklers (which are used to concentrate water at one point or to spread it over a large area). Hoses are usually attached to a hose spigot or tap.

Garden hoses are typically made of extruded synthetic rubber or soft plastic, often reinforced with an internal web of fibers. As a result of these materials, garden hoses are flexible and their smooth exterior facilitates pulling them past trees, posts and other obstacles. Garden hoses are also generally tough enough to survive scraping on rocks and being stepped on without damage or leaking.

Each male end of a typical garden hose can mate with the female connector on another, which allows multiple garden hoses to be linked end-to-end to increase their overall length. Small rubber or plastic washers (often confusingly called "hose washers") are used in female ends to prevent leakage, because the threads are not tapered and are not used to create a seal. Sometimes the gaskets stiffen, disintegrate, or fall out of older hoses, which results in pressurized leakage spraying from the hose; simply replacing the washer insert often fixes the problem.

Most garden hoses are not rated for use with hot water, and their packaging will often specify whether or not this is the case. Leaving non-reinforced hoses in the hot sun while pressurized can cause them to burst.

Hoses used to carry potable water are typically made of NSF International-listed polymers tested and shown not to leach harmful materials into the drinking water, such as the plasticizers (phthalates) used in polyvinyl chloride (PVC, or vinyl) hoses.

Usage

As implied by the name, garden hoses are commonly used to transport water for gardening, lawn care, and other landscaping purposes. They are also used for outdoor cleaning of items such as vehicles, equipment, building exteriors, and animals. NSF-approved hoses may be used for connecting drinkable water to recreational vehicles and trailers.

Sprayer pistol uses a quick-connect fitting, visible just beyond the sprayer grip.

Whenever a flexible hose is connected to a drinkable water supply, the spigot or tap should be fitted with an approved backflow prevention device, to prevent contaminated water from being siphoned back, in the event of a pressure drop. Many water suppliers require this, and plumbing code may legally require permanently installed backflow preventers.

Porous or Perforated Hoses

Special hoses designed to leak throughout their length are sometimes used to gently distribute water on a lawn or garden. These hoses either have many small holes drilled or punched in them, or are deliberately formulated of a porous material, such as sintered rubber particles. These "soaker hoses" are a simple, low-cost, crude type of drip irrigation system.

Expandable Hoses

These differ from traditional hoses in that the inner membrane expands when filled with water, much like a balloon. An outer cover protects the delicate expandable membrane from punctures. Such hoses "grow" when pressurized, and shrink back down when the pressure is released, allowing for easier storage.

Standards and Connectors

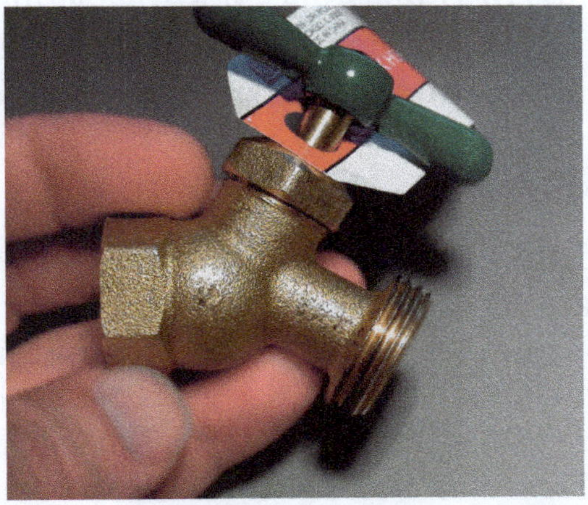

Brass hose spigot with standard tapered pipe threads (internal, not visible) and GHT (visible) threads

Garden hoses connect using a male/female thread connection. The technical term for this arrangement is a "hose union". Spigots or sillcocks have male hose connectors only, and the mating end of a hose has a captive nut, which fits the threads there.

The thread standard for garden hose connectors in the United States, its territories, and Canada is known colloquially as "garden hose thread" (GHT), but its official designation is NH (NH stands for "National Hose"; 3⁄4-11.5NH is for full form threads as produced by cutting material such as the brass spigot outlet or hose male or female end fitting found on more expensive hoses; 3⁄4-11.5NHR is for thin-walled couplers produced by rolling thin material, usually brass, typically found on less expensive hoses; 3⁄4-14NPSH is for female hose ends that mate a hose to a tapered pipe thread without a spigot). The standard was defined by NFPA 1963, "Standard for Fire Hose Connections", then later by ANSI-ASME B1.20.7, which is 1 $^{1}/_{16}$ inches (27 mm) diameter straight (non-tapered) thread with a pitch of 11.5 threads per inch (TPI). The female thread is abbreviated FHT, and the male part is abbreviated MHT. This fitting is used with $^{1}/_{2}$-inch, $^{5}/_{8}$-inch, and $^{3}/_{4}$-inch hoses.

In other countries, a British Standard Pipe (BSP) thread is used, which is $\frac{3}{4}$ inch and 14 TPI (male part outside diameter is 26.441 mm or 1.04 in). The GHT and BSP standards are not compatible, and attempting to connect a GHT hose to a BSP fitting, or vice versa, will damage the threads.

Various adaptors made of metal or plastic are available to interconnect GHT, BSP, NPT, hose barb, and quick connect fittings.

Quick Connectors

Starting in the 1990s, the use of quick-connector systems has become more popular. These are fittings that screw into the common hose connectors, allowing hoses and accessories to be easily connected together using a snap-fit system. The style manufactured by Gardena is common, imitated by and compatible with many other manufacturers. The connectors may optionally include an internal valve that is only opened by connecting the fitting, so that disconnecting a hose using this adaptor causes the water flow to stop. This greatly eases common tasks by allowing specialized sprayers to be interchanged without requiring two trips back to the spigot for each change.

Health Risks from Aerosols

In 2014, it was reported that use of common garden hoses in combination with spray nozzles may generate aerosols containing droplets smaller than 10 μm, which can be inhaled by nearby people. Water stagnating in a hose between uses, especially when warmed by the sun, can host the growth and interaction of *Legionella* and free-living amoebae (FLA) as biofilms on the inner surface of the hose. Clinical cases of Legionnaires' disease or Pontiac fever have been found to be associated with inhalation of garden hose aerosols containing *Legionella* bacteria. The report provided measured microbial densities resulting from controlled hose conditions in order to quantify the human health risks. The densities of *Legionella spp.* identified in two types of hoses were found to be similar to those reported during legionellosis outbreaks from other causes. It was proposed that the risk could be mitigated by draining hoses after use.

Garden hose in use.

Closeup of a garden hose shows the crisscrossed fiber reinforcements.

Gardena quick-connect hose fittings.

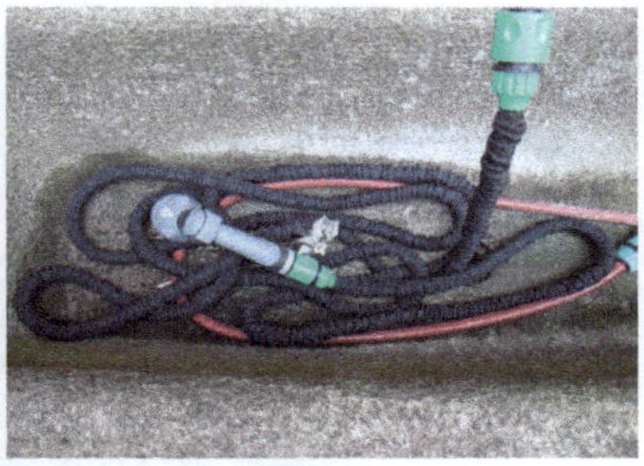

Special expandable hose can be stretched to reach farther.

Hose cart and some quick-connect fittings.

This greenhouse is equipped with a hose suspended on pulleys.

Subirrigation

Subirrigation is a greenhouse irrigation method that relies on capillary action to provide plants with water and nutrients from below their containers. The first documented subirrigation system was described in 1895, and several variations on the basic design were used for research purposes before the modern ebb-and-flow type systems emerged in 1974. Most subirrigation systems apply the fertilizer solution to a waterproof bench or greenhouse section, allowing the substrate to absorb the water through holes in the bottom of the containers. Because there is little or no leaching, subirrigation typically allows for the use of lower fertilizer solution concentrations. Although excess fertilizer salts typically accumulate in the top layer of the substrate, this does not seem to have a negative impact on plants. Subirrigation can conserve nutrients and water, reduce labor costs, and help growers meet environmental regulations. A challenge with subirrigation is the potential spread of pathogens via the fertilizer solution. When this is a concern, effective disinfection methods such as ultraviolet radiation, chlorine, or ozone should be used. Sensor-based irrigation control has recently been applied to subirrigation to further improve nutrient and water use efficiencies. Better control of irrigation may help reduce the spread of pathogens, while at the same time improving crop quality. The primary economic benefit of subirrigation is the reduction in labor costs, which is the greatest expenditure for many growers.

Greenhouse production is a vital part of the horticulture industry, with nearly 20,000 acres devoted to commercial greenhouse production in the United States alone. Frequent fertilization and irrigation is necessary, because plants typically grow in a small volume of substrate from which nutrients and water are rapidly depleted. Common greenhouse practices include overhead and drip irrigation, and plants are typically watered according to a set schedule. When more water is applied than can be absorbed by the substrate, excess water and nutrients leach. This leachate can be collected and reused, but often runs off to the external environment. Nutrients lost through runoff can contribute to nitrate and phosphate contamination of ground and surface water. Government regulations gradually require greenhouse growers to minimize their environmental impact by limiting nutrient runoff. Increasing fertilizer costs provide further incentive for growers to conserve nutrients and prevent losses.

Subirrigation is an irrigation technique that provides water or fertilizer solution to the bottom of containers. Capillary action of the substrate provides roots with water and nutrients. Greenhouse sections or benches with container-grown plants are periodically flooded within a closed system. The water or fertilizer solution is absorbed by the substrate through holes in the bottom of the container. The amount of water absorbed depends on substrate dryness, and irrigation volume closely matches plant requirements. Excess fertilizer solution is collected and reused for subsequent irrigation events, eliminating the environmental consequences associated with discharge, as well as reducing fertilizer costs for the producer. Subirrigation also reduces the cost of labor, which is the largest expenditure for horticultural producers.

Subirrigation has been used since at least 1895, when it was described by researchers at the Ohio Experimental Station. The subsequent development of hydroponics during the 1920s and sand-culture techniques in the 1930s helped establish the principles underlying modern subirrigation systems. Independently conceived subirrigation systems were also in use at the New Jersey and Purdue University agricultural experiment stations during the 1930s.

An early sand-culture system used two separate tanks to supply irrigation water and collected runoff from a sand bed. Excess water was eventually returned to the supply tank and electrical conductivity (EC) measurements were used to monitor nutrient concentrations of the recirculated solution. A later, more refined system provided uniform nutrient supply, increased aeration, and further minimized waste. In this nutrient-culture system, sealed benches containing an inert medium were periodically flooded with fertilizer solution, which was collected for reuse in the same reservoir used to flood the benches. Timers were eventually added to reduce labor costs. Other adaptations of these basic designs were developed and used for research and commercial production through the 1930s and 1940s. In all of these systems, plants grew directly in irrigated sand or other substrate, rather than in containers.

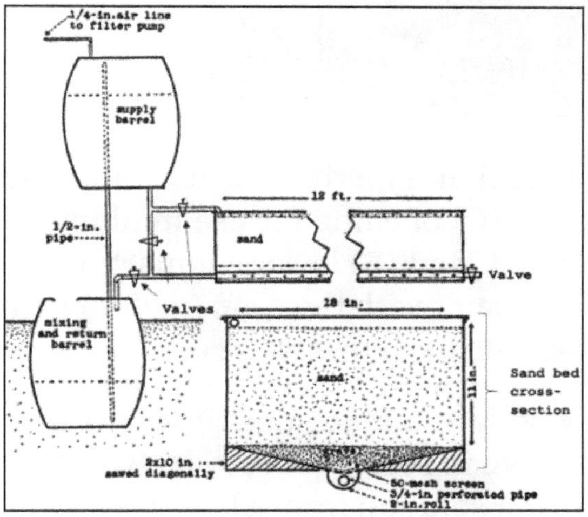

Section of sand-culture apparatus showing the sand bed, solution barrels, and plumbing; 1 inch (in.) = 2.54 cm, 1 ft (ft.) = 0.3048 m, and 50 mesh = 0.297 mm (0.0117 inch).

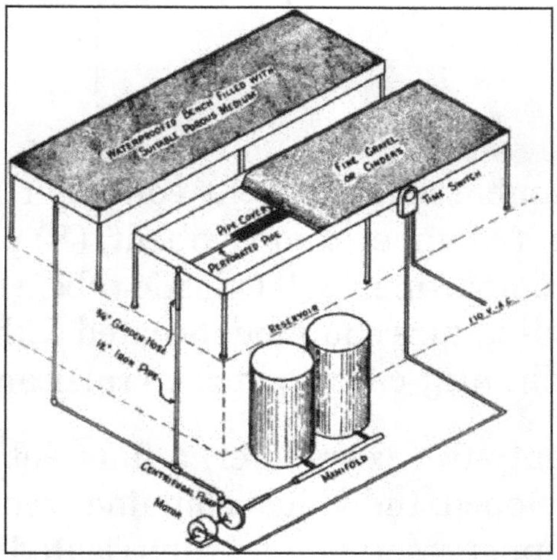

Subirrigation system for large-scale operation using two different substrates (porous media and fine gravel or cinders).

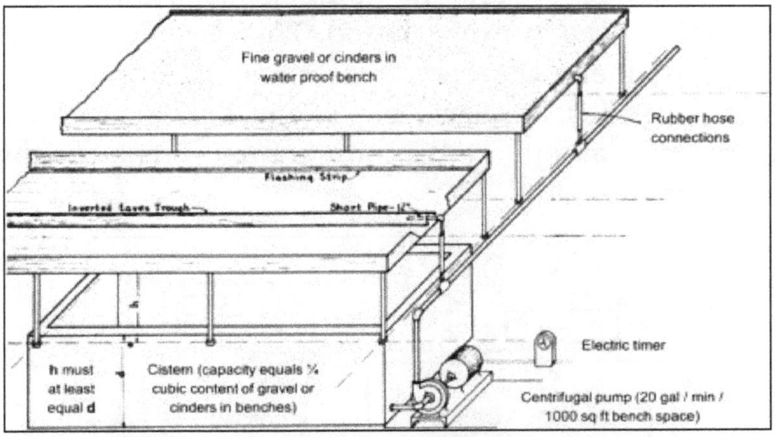

Subirrigation method of fertilizer solution culture.
At that time, all pipes were made by black iron.

Subirrigation with containerized plants was first described in 1950 as an improved and simplified alternative to the sand-culture and nutrient-culture techniques. In these systems, relatively small sand beds in one compartment were subirrigated with fertilizer solution from a lower reservoir. This system was used to commercially produce african violet (*Saintpaulia ionantha*) and impatiens (*Impatiens walleriana*), and was later redesigned to accommodate larger plants for experimental purposes .Unlike modern subirrigation systems, sand was used as the growing medium and fertilizer solution had to be discharged periodically due to changes in nutrient concentration.

A double compartment container for subirrigation experiments in plant nutrition. Left: (A) upper compartment, (B) plastic collar, (C) opening for use in filling the lower compartment, (D) plastic tube, (E) perforation, and (F) lower compartment. Right: Sand-culture equipment coated with aluminum paint with the 1-gal (3.78 L) bottle reservoir. A 15-cm (5.9 inches) ruler on the lower left side indicates size.

Twenty-two years later, a fully automated, highly versatile subirrigation system was developed. Individual containerized plants, grown in soilless substrate, were housed in an upper reservoir and subirrigated with fertilizer solution from a lower reservoir using a submersible pump activated by a timer. Unabsorbed fertilizer solution returned to the lower reservoir through a slow drain and could be reused indefinitely. The system drained more slowly than it was filled, thereby allowing sufficient time for uptake of fertilizer solution by capillary action, and maintaining substrate water content near container capacity throughout the growing period. This system could be easily modified for use with various container and reservoir sizes and configured so that one reservoir could supply multiple benches. Similar designs, now known as ebb-and-flow systems, subsequently became popular and are the most common type of subirrigation used in the greenhouse industry.

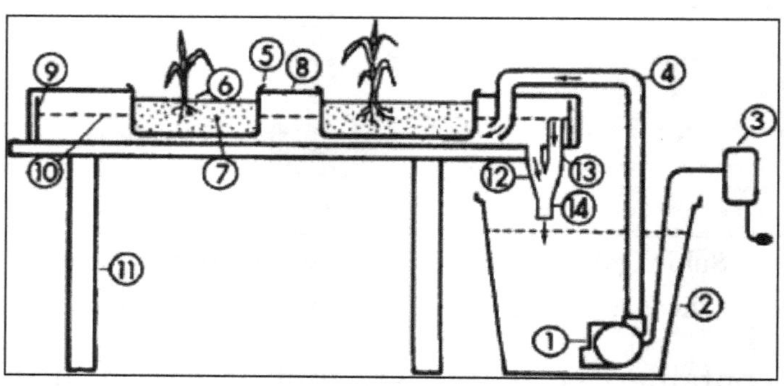

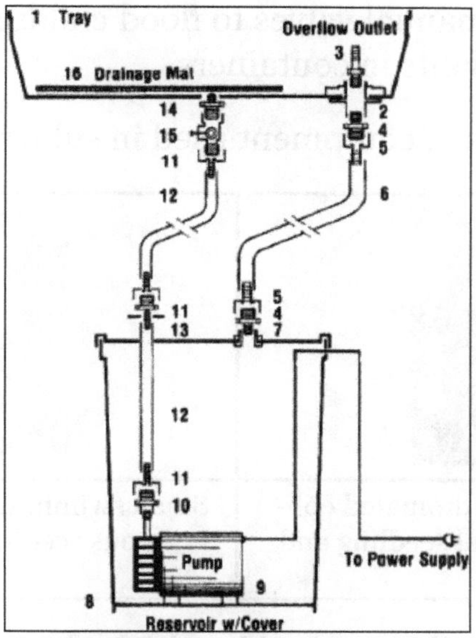

Pulsed subirrigation tray system.

(1) structural foam tray (0.36 × 0.50 × 0.10 m), (2) bulkhead fitting (1/2-inch dual thread), (3) adapter [1/2-inch male pipe thread (MPT) × 1/2-inch barb], (4) adapter [1/2-inch MPT × male garden hose thread (MHT)], (5) swivel [female garden hose thread (FHT) × 1/2-inch barb], (6) garden hose (1/2-inch), (7) bulkhead fitting (1/2-inch single thread), (8) bucket (18 L) with lid, (9) submersible pump with 1/4-inch MPT outlet, (10) adapter (1/4-inch female pipe thread × MHT), (11) swivel (FHT × 1/4-inch barb), (12) polyethylene tubing (black 1/4-inch), (13) adapter (MHT × 1/4-inch barb), (14) adapter (MHT × 3/8-inch MPT), (15) ball valve (1/4-inch turn), (16) drainage mat; 1 m = 3.2808 ft, 1 inch = 2.54 cm. 1 L = 0.2642 gal.

Equipment

Subirrigation systems are classified as ebb-and-flow benches, flood-floor, trough-tray, wick system, mobile or Dutch trays, and capillary mat. A study evaluating subirrigation use in 50 greenhouse establishments in 26 states indicated that ebb-and-flow is used by 58% of the growers, flood-floor by 13%, and trough-tray systems by 8%, with 21% of the interviewees having two or more systems.

The typical ebb-and-flow system includes an elevated, water-tight bench where plants are grown, a fertilizer solution reservoir, and a pump .The bench is periodically flooded with fertilizer solution pumped from the reservoir to which any unabsorbed fertilizer solution eventually returns through a gravity drain at a sufficiently slow rate to allow for absorption. Irrigation frequency can be controlled using timers. Dutch trays, also known as mobile trays, are self-contained mobile ebb-and-flow benches that can easily be moved throughout a greenhouse. These trays are manufactured specifically for use in highly automated growing operations. While manufactured ebb-and-flow systems are common in commercial greenhouses, similar systems can be easily assembled from inexpensive materials to suit specific needs, and may be designed to operate without a pump. For example, used manual valves to flood children's swimming pools for production of large deciduous plants in containers.

The figure below shows types of equipment used in subirrigation.

Commercial prefabricated and automated ebb-and-flow benches for ornamental seedling and plant production.	Small swimming pools adapted for large deciduous seedling production in containers.

Flood-floor irrigation is a similar method in which plants are placed directly onto the greenhouse floor, typically made of concrete, and the entire space is flooded through holes in the floor. The floor has a gentle slope to allow the water to drain back to the fertilizer solution tank. This maximizes the proportion of greenhouse space used for production and operates more rapidly, but is often more labor-intensive than standard ebb-and-flow bench systems. A relatively new approach to flood-floor irrigation is the partial saturation ebb-and-flow watering, whereby the fertilizer solution is pumped onto the higher side of a sloped floor and flows down to drain at the low end. This creates a thin film of water on the floor and allows growers more control over how much water the substrate absorbs. Compared with standard flood-floors, partial saturation systems use less water and fertilizer and produce smaller plants.

In the trough-tray system, plants are placed in sloped shallow gutters or troughs. Fertilizer solution is pumped into the higher side of the troughs and runs down to a drain at the lower side and returns to the fertilizer solution tank. Trough-trays are most useful for operations that produce continuously in the same size containers. They are less versatile than ebb-and-flow benches because they can accommodate only a limited number of plants in relatively small containers, and take up more space for a similar number of plants compared with ebb-and-flow benches because the narrow troughs must be spaced apart. However, trough trays also allow for better air circulation through the canopy because of this spacing.

Other types of subirrigation include wick irrigation and capillary mats. Wick irrigation supplies water and nutrients from a fertilizer solution reservoir to the substrate via an absorptive wick, providing consistent moisture without runoff. Capillary mats are absorbent mats used to provide potted plants with moisture from below the containers to minimize fluctuations in substrate water content. A finely perforated, thin plastic film can be used to cover capillary mats to reduce evaporation and algal growth. Newer capillary mats may also have a liner below the mat, which prevents water from dripping from the mat. Capillary mats are versatile and can be used on a temporary basis and for outdoor production.

Principles Underlying Subirrigation, Substrates, and Containers

Water and nutrients are delivered to plants by the passive movement of water through the substrate due to capillary action. Substrate physical properties may affect the efficiency of capillary rise. Adequate media stability, density, particle structure, and water holding capacity are needed to allow water movement within the containers. Substrates with large particles have large pore spaces, reducing capillary action. Most commonly used soilless substrates are suitable for use with subirrigation, and different substrates can be mixed to suit particular needs.

Container height can also affect subirrigation efficiency because water must travel further to reach the upper part of the substrate in taller containers have shown that vertical gradients occur in subirrigated containers because water is absorbed by the lower part of the substrate and moves upward. Capillary rise can occur slowly; in 15-cm-diameter × 12-cm-tall pots filled with a peat–perlite substrate, water did not reach the upper substrate layer until up to 20 h after 'Panama Red' hibiscus (*Hibiscus acetosella*) plants were subirrigated.

While plastic containers are most common, containers made from biodegradable materials can also be used with subirrigation. compared the effects of using several types of biocontainers for growing cyclamen (*Cyclamen persicum*) using ebb-and-flow benches over a 15-week production cycle. Shoot dry weight was greater in all biocontainers than in the plastic control pots, with the exception of wood fiber containers. Plants grown in the wood fiber containers had reduced dry weights, presumably because these containers did not have holes in the bottom and water could not easily be absorbed through the walls. These containers also absorbed less nutrient solution per irrigation and had the shortest irrigation interval. Highly porous biocontainers (made from wood fiber, peat, dairy manure, and rice straw) required a higher total volume of nutrient solution over the course of the study because water readily evaporated from the container walls. These containers also had greatly reduced tensile strengths at the end of the production cycle. Bioplastic, rice hull, paper, and coconut fiber containers did not differ from the control containers in irrigation interval or total irrigation volume, and did not have reduced tensile strengths at the end of the production cycle. compared ebb-and-flow subirrigation to hand and drip watering using coleus (*Solenostemon scutellarioides*) grown in several types of biocontainers. Subirrigation improved growth in all treatments, and this effect was attributed to increased fertilization rate due to the absence of leaching. Subirrigation reduced the puncture strength of manure, paper, and wood fiber biocontainers, but not peat biocontainers.

Water Conservation

Subirrigation can be especially beneficial in areas where water scarcity is an issue. Compared with overhead irrigation, subirrigation systems have been consistently shown to reduce overall water use, primarily because excess water is collected and reused. For example, found that subirrigation requires 56% less water than overhead irrigation. Roeber indicated the different water use for potted plant production, showing that ebb-and-flow benches and troughs use 0.4 to 0.8 $m^3 \cdot m^{-2}$ water per year, drip irrigation use 0.8 to

1.6 $m^3 \cdot m^{-2}$ water per year, and hand or sprinkler irrigation use 1.2 to 2.4 water $m^3 \cdot m^{-2}$ per year. Subirrigation generally uses less water than overhead watering methods mainly because the unused solution is collected for reuse, rather than being lost due to drainage.

Salinity

In semiarid regions, growers often rely on saline water sources. Subirrigation can mitigate the effects of osmotic stress that can be induced by applying saline water. Tomato (*Solanum lycopersicum*) subirrigated with fertilizer solution prepared with saline water had fruit yields comparable to plants drip irrigated with the same solution. Another study demonstrated that subirrigated tomato yields were higher and salt accumulation was minimal when nutrient content of the irrigation solution was reduced by 30%. Subirrigation reduced tomato plant size and overall water use but improved water use efficiency. Accumulated salts were found primarily in the upper substrate layers. The accumulation of salts within the substrate prevented the fertilizer solution from becoming overly saline, reducing the need to discharge unusable fertilizer solution, resulting in more efficient water and fertilizer use. In zucchini squash (*Cucurbita pepo*), subirrigation with moderately saline water reduced growth rate and yield, but improved fruit quality and water use efficiency when compared with subirrigation with nonsaline nutrient solution.

Vegetable and Fruit/Tree Crops

Currently, subirrigation is primarily used for ornamental plant production. However, subirrigation may become an increasingly valuable method for the production of non-ornamental crops, such as vegetables and fruit/tree seedlings.

Subirrigation type	Species	Summary
Ebb-and-flow benches (1 × 1 × 0.15 m)	Cyclamen (*Cyclamen persicum*)	Plants grown in several types of biocontaners had higher shoot dry weights than those grown in plastic pots
Ebb-and-flow troughs (5 m^2)	Chrysanthemum (*Dendranthema indicum*)	Used subirrigation to root chrysanthemum cuttings. Root growth was inversely correlated to substrate water content between waterings
Hand-watered plant saucers (0.195 m diameter)	Poinsettia (*Euphorbia pulcherrima*)	Subbirigation was compared with overhead watering. There was no difference in plant height or bract size, but subirrigated plants had lower dry weights and diameters
Ebb-and-flow benches (1.5 × 0.9 × 0.04 m)	'Panama Red' hibiscus (*Hibiscus acetosella*)	Subirrigaton was automated using soil moisture sensors
Potted plant system (dimensions not available)	Nephthylis (*Syngonoum podophyllym*)	Describes the design of a subirrigation system that operates without pumps
Hand-watered trays (0.5 × 0.35 × 0.06 m) lined with 6-mil black plastic	Rhododendron (*Rhododendron* sp.)	Hardwood stem cuttings had large root balls and a higher rooting percentage in lower pH media

Ebb-and-flow benches (2.4 × 1.2 m)	Petunia (*Petunia ×hybrida*) and wax begonia (*Begonia ×semperflorens-cultorum*)	Good growth and flowering over wide range of fertilizer solution concentrations, little effect of different substrates
Ebb-and-flow benches (2.4 × 1.2 m)	Petunia	Optimal fertilizer electrical conductivity (EC) decreases as temperature increases. Maintaining optimal leachate EC is better than maintaining constant fertilizer solution EC
Ebb-and-flow benches (2.4 × 1.2 m)	Alyssum (*Lobularia maritima*), celosia (*Celosia argentea*), dianthus (*Dianthus chinensis*), (*Gomphrena globosa*), stock (*Matthiola incana*), and zinnia (*Zinnia elegans*)	Optimal fertilizer concentrations were species-dependent. Superoptimal concentrations had little impact on growth
Ebb-and-flow benches (2.4 × 1.2 m)	Salvia (*Salvia splendens*)	High fertilizer concentrations increase the shoot:root ratio and hasten flowering
Ebb-and-flow benches (2.4 × 1.2 m)	Petunia and wax begonia	Optimal leachate EC is 1.0 to 1.7 $dS{\cdot}m^{-1}$
Ebb-and-flow trays (1 × 1.1 m)	New guinea impatiens (*Impatiens walleriana*) and peace lily (*Spathiphyllum wallisii*)	Growth was most vigorous at 8 mM nitrogen (N) for new guinea impatiens and 10 mM N for peace lily, and declined thereafter. EC was two to five times higher in the upper substrate in all treatments
Ebb-and-flow benches (dimensions not available)	Areca palm (*Dypsis lutescens*) and philodendron (*Philodendron* sp.)	Compared with overhead watering, subirrigation used 10 times less water per unit area
Ebb-and-flow benches (dimensions not available)	Petunia	Subirrigated plants grew larger than overhead-watered plants at several rates of controlled release fertilizer incorporation. With liquid fertilizer, subirrigated plants were larger at 100 ppm N, but smaller at 150 ppm N
Flood floor (5 × 9 m) with a slope of 0.5%	Fortune spindle (*Euonymus fortunei*) and northern white-cedar (*Thuja occidentalis*)	Subirrigated plants had higher shoot dry weights at the lower of two fertilization rates. Both species grew less than overhead-watered plants in peat
Ebb-and-flow (dimensions not available)	Shamrock (*Oxalis regnellii* and *Oxalis triangularis*)	Overhead-watered plants were larger than subirrigated plants across a range of fertilization rates
Ebb-and-flow benches (1.5 × 1.8 m) and capillary mats	Geranium (*Pelargonium ×hortorum*)	Subirrigation was compared with microemitters and hand watering. Overall plant growth was similar in all treatments, but root numbers were higher in the subirrigated plants
	Subirrigation for ornamental plant production	
Ebb-and-flow benches (2.4 × 1.2 m)	Wax begonia	Optimal fertilizer concentration was estimated from water use efficiency
Ebb-and-flow benches (2.4 × 1.2 m)	Wax begonia and petunia	Light level did not affect optimal fertilizer solution concentration
Ebb-and-flow benches (2.4 × 1.2 m)	English ivy (*Hedera helix*)	Marketable plants were grown at the lowest fertilization rate (100 ppm N) in three light treatments

Ebb-and-flow benches (2.4 × 1.2 m)	Pale purple coneflower (*Echinacea pallida*)	Subirrigated plants had higher survival rates, greater height and biomass, and better N use efficiency than overhead-watered plants
Ebb-and-flow light weight 1.486 m^2 plastic benches	Croton (*Codiaeum variegatum*), dieffenbachia (*Dieffenbachia maculata*), and spathiphyllum (*Spathiphyllum ×petite*)	Plants grew larger in commercial potting mixes than in a custom peat:pine bark (1:1) mix. There was little response when N fertilization rate was increased above 4.2 $g{\cdot}L^{-1}$
Trough-tray (5 × 0.16 m)	Geranium	Compared with a recirculating drip system, subirrigation led to higher substrate EC and less variation in nutrient solution (NS) EC. Subirrigated plants grew poorly at full strength NS
Ebb-and-flow (dimensions not available)	New guinea impatiens	Plant growth and quality decreased as substrate salinity increased. EC was highest in the upper substrate across treatments
Ebb-and-flow benches (0.9 × 1.5 m^2)	Poinsettia	Post-production top watering can wash salts into root zone, but is not likely to cause damage
Capillary mat (1.8 × 1.8 m)	Pansy (*Viola ×wittrockiana*)	High fertilizer rates increase growth and shoot:root ratio
Capillary mat (1.8 × 1.8 m)	Marigold (*Tagetes patula*)	Drought stress reduces growth and plant height, but does not increase compactness
Hand-watered trays (0.51 × 0.13 × 0.06 m)	Chrysanthemum (*Dendranthema ×grandiflorum*), coleus (*Solenostemon scutellarioides*), red maple (*Acer rubrum*), and japanese tree lilac (*Syringa reticulata*)	Subirrigation can effectively replace misting for rooting cuttings of some ornamental crops
	Subirrigation for vegetable production	

Subirrigation type	Species	Summary
Hand-watered trays (0.51 × 0.13 × 0.06 m)	Lamarck serviceberry (*Amelanchier lamarckii*), amur maackia (*Maackia amurensis*), cherry (*Prunus serrulata*), spirea (*Spiraea ×bumalda*), lilac (*Syringa vulgaris*), and elm (*Ulmus glabra × Ulmus carpinijolia*)	Subirrigation without mist was effective for rooting softwood cuttings of elm, lilac, and spirea
Prototype subirrigation trays (0.62 × 0.41 × 0.15 m)	Rangpur lime (*Citrus limonia*)	Subirrigation induced higher plant growth and shortened the crop cycle compared with hand-water irrigation, anticipating the transplanting to the next phase (grafting)
Ebb-and-flow trays (1.22 × 1.22 m)	Northern red oak (*Quercus rubra*)	Under fertilized conditions, subirrigated seedlings had greater biomass and field diameter growth than top-watered plants
Subirrigation (dimensions not available)	Eucalyptus (*Eucalyptus grandis*, *Eucalyptus urophylla*, and *E. grandis × E. urophylla*)	Subirrigation increased the number of cuttings on clonal plants

Ebb-and-flow benches (2.44 × 1.22 × 0.115 m)	Koa (*Acacia koa*)	Subirrigated and overhead-watered seedlings had similar growth and vigor
Subirrigation system (3.75 × 0.85 m) with expanded clay	Coffee (*Coffea arabica*)	Adequate plant growth, regardless of the fertilization level of the stock plant
Subirrigation trays (0.62 × 0.41 × 0.15 m)	Rangpur lime	Validation of subirrigation technology for citrus rootstock production
Subirrigation system (dimensions not available)	Trifoliate orange (*Poncirus trifoliata*), citrange (*P. trifoliata* × *Citrus sinensis*), mandarin (*Citrus sunki*), and citrumelo (*P. trifoliata* × *Citrus paradisi*)	Different substrate composition influence on the capillary action
Subirrigation benches (1.65 × 2.55 × 0.2 m)	Araucaria (*Araucaria angustifolia*), jeriva (*Syagrus romanzoffiana*), cutieira (*Joannesia princeps*), mutamba (*Guazuma ulmifolia*), red angico (*Anadenanthera macrocarpa*), and pink peroba (*Aspidosperma polyneuron*)	Subirrigation showed high water use efficiency and good plant uniformity in conic containers

Table: Selected publications related to subirrigation for fruit/tree seedling production.

Fertilization

Recirculation allows overall fertilizer use to be reduced because no nutrients are lost from the system. However, subirrigation requires careful management of fertilizer solution concentrations to produce high-quality greenhouse crops. Optimal fertilization rates for overhead irrigation systems are well known, but there is less applied information available about ideal fertilizer solution concentrations for subirrigation. Generally, fertilizer concentrations should be lower with subirrigation than with overhead or drip irrigation. Nutrient salts are not leached from the substrate and can accumulate within the containers, potentially exposing the plants to osmotic stress. High salinity occurs mainly in the upper substrate layer because salts move along with the capillary flow. As water evaporates from the surface, the salts accumulate in the top substrate layer. The accumulation of salts in the upper substrate layer is exacerbated by high fertilization rates. However, subirrigated plants are generally unaffected by high salinity in the upper substrate layers because root growth occurs primarily in the lower portions of the container where there is more water available. In the postproduction environment, accumulated salts from the upper portion of the substrate can potentially be washed to the bottom layers by top watering, but this does not seem to cause serious damage to the plants. Effective nutrient management for subirrigation requires minimizing the risk of osmotic stress, while providing the plants with adequate nutrition.

Optimal fertilizer solution concentrations vary among species and may depend on both the nutritional requirements and salt tolerance of a particular crop. This was demonstrated in a comparison of several bedding plant species subirrigated with various concentrations of Hoagland solution (12.5% to 200%). Full strength (100% concentration) Hoagland solution has a nitrogen concentration of 210 $mg \cdot L^{-1}$ and an EC of 2.0 $dS \cdot m^{-1}$. Maximum zinnia (*Zinnia elegans*) and celosia (*Celosia argentea*) dry weights were observed at a 50% strength solution, and dry weights, as well as zinnia flower diameter, decreased at higher concentrations. In alyssum (*Lobularia maritima*) and dianthus (*Dianthus chinensis*), maximum growth occurred at the 100% concentration, but dianthus had the most flowers at a 200% concentration. Gomphrena (*Gomphrena globosa*) and stock (*Matthiola incana*) grew best within the 100% to 200% range, suggesting that these plants are particularly tolerant of accumulated salts . In a similar experiment, salvia (*Salvia splendens*) growth increased as fertilizer concentration increased from 12.5% to 100% Hoagland solution. Shoot dry weight, shoot:root ratio, and leaf area increased, while net photosynthesis, stomatal conductance, and transpiration decreased at higher fertilizer solution concentrations, suggesting that the treatment effects were due to shifts in carbon allocation and more efficient production of leaf area at increased fertilization rates. Leaf area decreased when fertilizer concentration was increased to 200%. Comparable effects of fertilizer solution concentration on plant growth were also observed in subirrigated pansy [*Viola×wittrockiana* (and wax begonia [*Begonia ×semperflorens-cultorum*]. However, found that fertilizer solution concentration had no effect on leaf area of subirrigated gerbera (*Gerbera jamesonii)* during the final production stage. Growth decreased when subirrigation fertilizer concentrations exceeded an optimum concentration in subirrigated poinsettia [*Euphorbia pulcherrima*], new guinea impatiens [*Impatiens ×hawkeri*], wax begonia , petunia [*Petunia ×hybrida*], miniature rose [*Rosa chinensis minima* ; collards (*Brassica oleracea*var. acephala), kale (*B. oleracea* var. acephala), lettuce (*Lactuca sativa*), pepper (*Capsicum annuum*), and tomato transplants ; and shamrock species [*Oxalis regnellii* and *O. triangularis*].

Agrochemicals

Subirrigation allows pesticides and plant growth regulators (PGRs) to be applied to plants along with the fertilizer solution. Such applications should only be made if the label specifically allows for application by subirrigation. Using subirrigation to apply pesticides and PGRs ensures uniform application, prevents the release of these chemicals to the environment, and can further reduce labor costs as well as minimize employee exposure to pesticides. A potential problem with applying agrochemicals via subirrigation is that excess solution drains back into the holding tank. Contamination can be prevented by using separate holding tanks for agrochemicals, but this may be expensive. However, residual concentrations of PGRs in fertilizer solutions, such as paclobutrazol and uniconazole, were shown to be very low after application via subirrigation. Another consideration in applying agrochemicals through subirrigation is the substrate moisture level: drier substrate absorbs more solution, and thus agrochemical, than wetter substrates. Subirrigation has also been used for applying natural PGRs such as tea seed (*Camellia* sp.) powder.

Imidacloprid (Marathon 60 WP; OHP, Mainland, PA) applied via subirrigation to poinsettias was more effective at controlling silverleaf whiteflies (*Bemisia argentifolii*) than drench applications. Although drench applications resulted in faster initial uptake of the imidacloprid, subirrigation application resulted in higher leaf concentrations of imidacloprid after 63 d and better long-term control. Applications by subirrigation appeared to allow for plant uptake of imidacloprid over a longer period, resulting in more uniform distribution of the insecticide throughout the plant, as compared with drip irrigated plants treated with an imidacloprid drench.

Pathology

Transmission of root-infesting pathogens between containers or benches through recirculated irrigation water is a potential drawback of subirrigation. Oomycetes of the genera *Pythium* and *Phytophthora* are particularly problematic, since they produce large numbers of highly mobile aquatic zoospores and seriously affect plant growth and quality. Bacteria, viruses, and fungal pathogens can also migrate through the recirculating system and infect new hosts. Several measures can be taken to reduce the pathogen load of recirculated water. Infectious propagules can easily enter the reservoir with plant debris, and removal of dead material helps minimize disease spread. Filtering the fertilizer solution can also prevent contamination, removing extant microbes and preventing future infections. Membrane filters effectively remove pathogens, but are expensive and require frequent replacement. Slow media filtration is the passing of recirculated irrigation water through an inert media, most commonly sand, at a low flow rate. This process eliminates most plant pathogens through a combination of mechanical filtration and biological activity, but may be impractical because it requires a large amount of space, may not consistently provide adequate filtration, is relatively slow, and can support microbial populations that include human pathogens. Recirculated fertilizer solution can also be treated using ozone or ultraviolet radiation to eliminate pathogens. However, in a survey of greenhouses and nurseries, Meador et al. found that recirculating fertilizer solution from subirrigation systems, on average, did not meet recommended standards for horticultural water quality, and that microbial counts were exceptionally high. The spread of disease among containers in subirrigation systems is an important practical concern and more effective and economical preventative methods are needed.

Limiting the duration of flooding events or the amount of fertilizer solution provided to subirrigated plants may prevent or minimize disease spread. Partial saturation ebb-and-flow watering rapidly delivers water to plants. With this method, less water is absorbed by the substrate per flooding event. Elmer et al. demonstrated that partial saturation can prevent the spread of pythium root rot (*Pythium* sp.) infections. This may be because less fertilizer solution drains out of the containers following irrigation, fewer pathogens survive in the drier substrate, or partial saturation may provide plant root zone with more oxygen.

Sensor-based Subirrigation Control

Subirrigation is typically controlled using timers according to a predetermined schedule, usually designed to meet operational needs. Sensors can be used to monitor substrate moisture and control subirrigation based on plant water use. For a review of the use of different sensors in irrigation control, . Subirrigation has been successfully automated using tensiometers , lysimeters , and capacitance moisture sensors . Electronic switches connected to tensiometers to begin and end irrigation events at −5 and −1 kPa substrate matric potential (high and low media tension values, respectively) in drip and trough-tray irrigated containers were tested in several studies . Subirrigation with a standard nutrient solution decreased growth rate and yield of zucchini squash during the spring–summer growing season, but not during the summer–fall growing season compared with the automated drip system . When squash plants were irrigated with fertilizer solutions made with saline water, tensiometer-controlled subirrigation decreased yield, but improved fruit quality and increased water use efficiency . Another study used this automation with nonsaline water and showed that tensiometer-controlled subirrigation resulted in zucchini squash fruit yields equal to drip irrigation when full-strength fertilizer solution was used. However, use of half-strength fertilizer solutions reduced yield more with subirrigation than with drip irrigation. With zonal geranium (*Pelargonium×hortorum*) grown in the spring and fall at half- and full-strength fertilizer solutions, subirrigation only reduced growth when full-strength solution was used during the spring. In a similar comparison, found that drip irrigated plants used more water and were larger than plants grown in trough-trays when irrigation was automatically triggered at −7 kPa. However, when relatively low fertilizer concentrations were used, subirrigation resulted in equal yields as open-cycle drip irrigation. Triggering irrigation based on specific substrate moisture levels cannot only reduce water use; it can also be used to control plant vigor, potentially reducing the need for plant growth retardant applications. Inexpensive open-source microcontrollers can be used to build low-cost automated irrigation controllers.

Economic Benefits of Subirrigation

While fertilizer and water use can be reduced with subirrigation, this alone may not be sufficient to offset the initial price of installing an automated subirrigation system, which is often high. The primary economic benefit of subirrigation is that, by facilitating automation, it reduces the cost of labor, which is the greatest expenditure for many producers. Compared with traditional overhead watering, subirrigation may also improve plant health and quality. The incidence and spread of foliar disease is reduced because the leaves are not wetted during irrigation. Furthermore, subirrigated crops are typically uniform, because water and nutrients are evenly distributed, and uniformity makes handling and shipping easier.

Manual Irrigation

Manual irrigation systems are easy to handle, require no technical equipment and are therefore generally cheap (in contrast to high-tech systems such as sprinkler irrigation or subsurface drip irrigation). But they need high labor inputs. A common and very simple technique for manual irrigation is for instance the use of watering cans as it can be found in peri-urban agriculture around large cities in some African countries. A more sophisticated and very water-efficient type of manual irrigation system is small-scale drip irrigation with buckets Beside these systems, there are many other methods for manual irrigation, which are easy to install and simple to use. In general, all of these methods have high self-help compatibility and a relatively high performance. Therefore such systems are also called helpful irrigation methods: high frequency, efficient, low-volume, partial-area, farm-unit, and low-cost.

Basic Manual Irrigation Principles

Low-Cost Drip Irrigation System

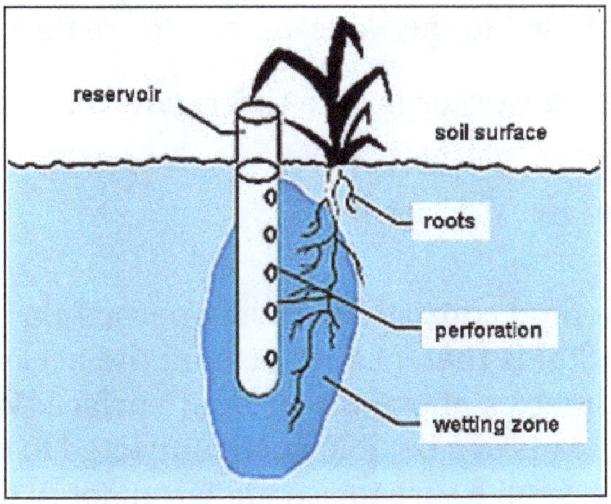

Simple drip irrigation (in contrary to high tech drip irrigation systems) uses low-cost.

plastic pipes cut to the appropriate lengths laid on the ground to irrigate vegetables, field crops and orchards. Small holes in the hose allow water to drip out and keep the base of the plant wet without wasting any water.

A low-cost "farm-kit system" with a 1000 litres water tank can service up one-eight of an acre.

Watering Cans

Irrigation by watering cans is a very basic way but is still widely used. This creates a lot of work for the labors especially if this technique is used for large fields. A common way to make this work easier is a carry-pole across the shoulders. The field worker is able to carry two big watering cans in each side and the irrigation water can de distributed equally on the field. A rose can be added to the watering-can to create a "sprinkler effect".

With watering cans, the field worker is able to irrigate very specific and only where it is necessary.

Pitcher Irrigation

A very basic subsurface method consists in placing porous clay jars (or pots) in shallow pits dug for this purpose. Soil is then packed around the necks of the jars so that their rims protrude a few centimetres above the ground surface. Water is poured into the jars either by hand or by means of a flexible hose connected to a water source. Since the walls of the pots are porous (make sure to use unglazed pots), the water can seep slowly out and reach the roots of the plants. The jars can be made of locally available clay: they are of no standard shape, size, wall thickness or porosity. Instead of a clay or earthenware pod, also the sweet monkey orange fruit (Strychnos spinosa) can be used when it has been dried and the top cut off. The jar should be filled up regularly (especially in arid areas) and has to be changed if there are big cracks that the water percolates without reaching the roots.

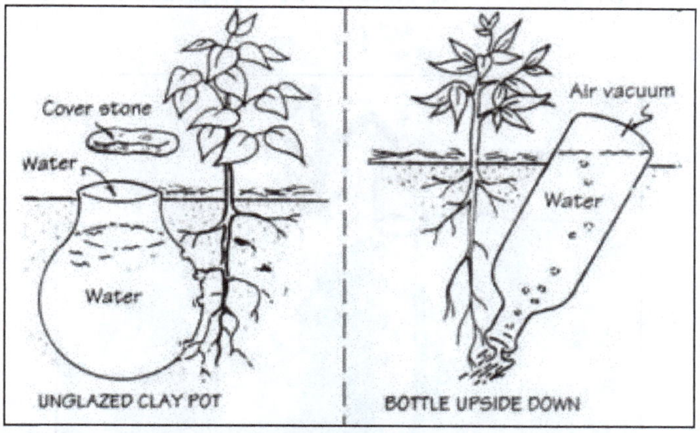

Left: Clay pot irrigation method. Right: Bottled irrigation method is also effective and simple. Bottles can be found everywhere in the world.

Bottle Irrigation

Similar to the pot method described above, pitcher irrigation can also be done using a bottle. The bottle is first filled and than placed with its neck into the soil next to a plant, so it stands upside down. The dense soil hinders the water from leaving the bottle immediately. Instead, it gets released slowly and directly besides the roots, so it is available to the plant for a longer time and the water cannot evaporate directly.

Porous and Sectioned Pipes

Another variation of pitcher irrigation uses porous pipes instead of pots to spread water along a continuous horizontal band in the soil, rather than at discrete locations. As such, the porous pipe method is more suitable for closely spaced row crops grown in beds, such as vegetable crops. One end of the porous pipe is made protrude above ground that the operator is able to refill it with water as soon as it is necessary. In contrast to subsurface drip irrigation, the porous pipe distributes the water over the whole length and not only where it is perforated. That means it is less effective and water loss is slightly higher.

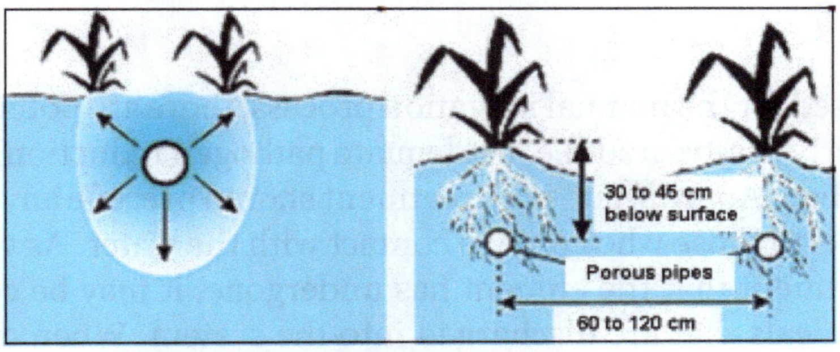

Schematic view of porous pipe irrigation.

Perforated Plastic Sleeves

An interesting variation of the pitcher method is the use of thin plastic sheeting to form a sleeve-like casing. To define its comparative usefulness better, the method should be tested side by side with alternative methods of irrigation. To date, this has not been done systematically.

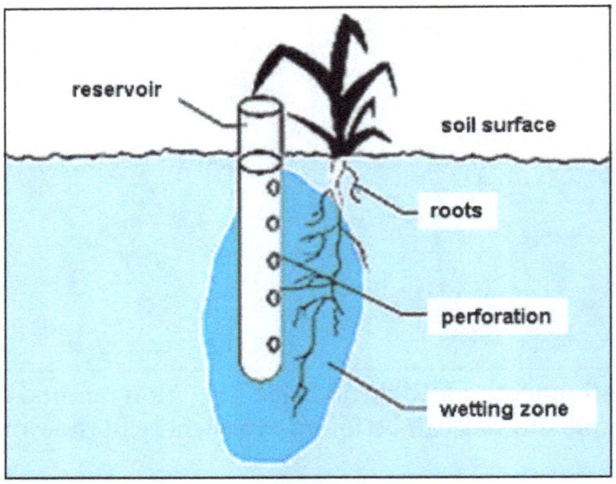

The plastic sleeve method is not tested systematically and therefore it is difficult to estimate its performance.

Costs Considerations

All of the described systems are very cheap. Many of them can be made out easily available material (e.g. old buckets or bottles) or renewable resources (bamboo). This reduces the costs significantly. However, high labour inputs are required for operation and maintenance.

Operation and Maintenance

As water is brought into the system manually, this requires high labour input. Moreover, it is important to check the systems regularly to prevent blockages and leakages. If there are any problems it should be cleaned and/or fixed as fast as possible to prevent damages on crops. Furthermore, there are several techniques to improve the production and avoid water loss on the fields.

Health Aspects

If wastewater is used for the manual irrigation process, there are potential health risks if water is not properly pre-treated (i.e. inadequate pathogen reduction). If poorly treated wastewater is applied. Appropriate pre-treatment should precede any irrigation scheme to limit health risks to those who come in contact with the water. As well, depending on the degree of treatment that the effluent has undergone, it may be contaminated with the different chemicals that are discharged into the system. When effluent is used for irrigation, households and industries connected to the system should be made aware of the products that are and are not appropriate for discharging into the system. Drip irrigation and subsurface drip irrigation are the only types of irrigation that should be used with edible crops, and even then, care should be taken to prevent workers and harvested crops from coming in contact with the treated effluent. Despite safety concerns, irrigation with effluent is an effective way to recycle nutrients and water.

Applicability

Manual irrigation methods are appropriate for small-scale farming or backyard gardening irrigation in dry and arid climates where water is scarce. All the different designs reduce water evaporation. It allows people to grow its own food with simple but effective techniques.

Advantages	Disadvantages
Improved water-use efficiency (reduced loss through evaporation)	Labour intensive
Well directed, selective and targeted irrigation	User need a basic training to install and use the correct most of the method
Ensures constant water supply in the crucial phase of germination	If the water is not properly filtered and the equipment not properly maintained, it can result in clogging
Higher yields, better quality, higher germination rate, lower incidence of pest attack	Manual subsurface drip irrigation avoids the high capillary potential of traditional surface applied irrigation, which can draw salt deposits up from deposits below
Facilitates pre-monsoon sowing	
Can be constructed with locally available material	
Low investment costs	

Gravity Irrigation

Gravity irrigation is a method of irrigation in which you can water plants anywhere, in.

the ground or in containers, even if there's no electricity available. These systems work by taking advantage of water's ability to travel the path of least resistance all on its own. You don't need any pumps and can use this for watering your garden temporarily or all season long. It is best used on individual plants, small clusters, or row crops.

Without a pump, you need a reservoir. You can fill it with collected rainwater, from the hose, or transported from nearby a pond or stream. The water moves to your plants' roots through drip irrigation. Just don't forget to check your water level. It will run out faster than you expect during a drought or a dry climate, especially as your plants grow larger. You can also add fertilizer or nutrient supplements to your irrigation reservoir.

Since delivery of gravity irrigation is via drip rather than spray, it conserves water. And if constructed right, you can use it for sub-irrigation, reducing evaporation loss even more. However, below the surface drip irrigation might call for a reservoir for every plant. It depends on the size of your plants and your reservoir.

You can buy gravity irrigation kits and systems, but it's easily turned into a DIY irrigation project. You can use water or soda bottles for individual plants. To water a row or group of plants from one drip irrigation reservoir it's best done using buckets, totes, or barrels with a tight-fitting lid. You don't want to lose water to evaporation from the container. It's also important to stop debris from collecting that can plug up your drip system.

One ancient gravity irrigation system is known as an Olla pot. These clay vessels buried in the middle of a ring of plants drip water through the porous sides whenever the water pressure around it drops too low to hold it in. You can use one alone or create a connected system that automatically refills a chain of them from a reservoir. Garden shops sell the clay versions, but a plastic drink bottle works well too. Drill tiny holes down one or all sides and bury it next to plant roots. Using the cap end for refills works better than cutting off the bottom. Gardeners use them in flower pots, container gardens, and traditional plots. However, you need a larger reservoir for large plants and multiple plant like a big jug.

Setting up gravity irrigation that delivers water at the surface calls for elevating your reservoir. The simplest setup makes use of 1-2 liter PET bottles with either hole drilled in the cap or purchasing spike emitter caps. Since this leaves your reservoir exposed to sunlight, it's best to use transparent bottles as a temporary solution while you're out of town. Because if the sun can penetrate it algae will grow in your reservoir. For long-term use, paint repurposed drink bottles a dark color.

Got an army of containers? A raised bed or backyard garden? Consolidating your DIY irrigation to draw from a larger drip irrigation reservoir might make far more sense. Here you will want to add a hose timer to manage water flow because, with the water reservoir raised off the ground, the flow won't shut off by pressure in the soil. And you do want it elevated to improve your water pressure unless you only need it to travel short distances.

One urban farming nurse's solution in a cucumber patch was to incorporate IV drip lines used in hospitals. This would allow some special individual plant treatment and will work in a pinch, but perhaps isn't suited to all water sources. Hose or tap timers aren't always expensive and come in manual, solar, and battery-powered models. Just connect it to the spigot tapped into your raised reservoir and attach the main water line feed to the timer outlet.

In such a DIY irrigation system you can connect both weeping drip hose or drip line tubing delivering water down your rows or grid. Using drip line tubing allows you to choose where water is emitted. So, you use water only on your plants, which increases your water conservation. You can also attach spaghetti line and drip emitter caps to increase water delivery to larger root systems. However, drip emitters and spaghetti line can easily clog from debris and particles in pond, stream, and well water. If any of those are your water source, look for drip system parts rated for dirty water.

Gravity irrigation isn't just for outdoor gardening. In fact, a system like the AutoPot found at indoor gardening shops uses gravity fed sub-irrigation made available to the plants through capillary action. Here a specially designed mechanism controls water flow, triggered only when the growing media is dry.

You will also see people using the pop bottle irrigation reservoir for indoor plants. Some partially or fully bury them next to the plant. Others make use of the drilled or spike emitter cap. And then there's the primed line drip trick. Filling a jar or jug with water and placing it raised above the plants. Next, create your gravity irrigation via a piece of tubing much like siphoning gas. Suck the water to the end and stick it in the growing media. It will only let water into the container when depleted water pressure allows it to do so. Note that you need small tubing or you'll have more than a drip action.

There's an automatic irrigation option that fits anyone's style of gardening, no matter where on Earth you live. Gravity irrigation can keep plants happy and healthy with a fine-tuned manufactured system, a simple conversion kit, or full-blown DIY irrigation for a big garden or farm that need multiple reservoirs. It's not always necessary to have fancy stuff to grow well. All your veggies care about is consistent moisture, lots of sunshine, and a balanced diet of quality nutrients.

Lift Irrigation

Lift irrigation is a method of irrigation in which water instead of being transported by natural flow (as in gravity-fed canal systems) requires external energy through animal, fuel based or electric power using pumps or other mechanical means. Treadle pumps, although an ancient method of lifting water for small heads have recently been modernized and used in a big way.

Lift irrigation schemes must accomplish two main tasks: first, to carry water by means of pumps from the water source to the main delivery chamber, which is situated at the top most point in the command area. Second, they must distribute this water to the field of the beneficiary farmers by means of a suitable and proper distribution system. The source is mainly groundwater, river streams, contour canals, ponds and lakes.

For a viable lift irrigation scheme, the requirements are constant water source should for the whole irrigation season at the site and the feasibility to lift water to the desired location. Different capacity pumps are required depending upon the duty point head, and discharge. The rising main may be of steel, concrete or any other suitable material. Lift irrigation schemes are useful where the target land is at higher level.

The advantage of lift irrigation is the minimal land acquisition problem and low water losses. The lift irrigation schemes are instrumental in stabilizing agriculture production particularly in the years of droughts and increase food production as water is available whenever it is required and thereby increase in income level.

Lift irrigation schemes are either individually owned or owned by a group of farmers in a cooperative mode. For successful functioning the lift irrigation schemes require appropriate technique, planning, designing and execution through knowledgeable technical person. Participation of beneficiaries is quite necessary. Unplanned development of lift irrigation systems have the potential to have its adverse impact on the groundwater levels, as has been the case in many south Asian countries in the recent years. Continuous drop in groundwater table is making the cost of running and maintenance of lift irrigation schemes more costly. Cooperative lift irrigation schemes have the potential to be participatory in development and management. Multistage submerged pump is at the heart of any irrigation system.

Lift Irrigation on Surface Water

Size and Location

The cultural command area should be restricted up to 20- 40 ha only under small lift irrigation schemes. Lift irrigation schemes on surface water source with command area more than 40 ha may be considered as medium / large schemes and the technical and financial appraisal may be carried out as per the guidelines pertaining to large lift irrigation schemes.

The scheme area should be as close to source of water as possible and land holdings are contiguous as far as possible from design optimization point of view.

Command Area / Cropping Pattern

The command area has to be demarcated and as far as possible to be adjoining. The cropping pattern has to be fixed and water requirement has to be estimated.

Design Discharge

The design discharge for small lift irrigation scheme should be assessed based on peak water requirements for envisaged cropping pattern and area of crops during kharif, rabi and summer seasons based on water requirement norms.

Water Availability

It is necessary to ascertain the availability position of water from sources like river, canal, ponds, tanks or any other surface water body.

It should preferably be perennial in nature and should have adequate flow to support the quantity of water to be pumped for the proposed scheme.

Water Lifting Permission

Water lifting permission should be obtained for private individuals or societies from Government Department authorized to issue such permissions. The water lifting permission should indicate the period, area and percentage of different crops in each season.

Long duration water lifting permission is required if the implementing agency desires to take loan.

Soil Characteristics

Soil characteristic should broadly be known for its suitability for various crops proposed in the scheme.

Power

Normally the power availability for large schemes is considered to be 16 hours per day but in case of small lift irrigation scheme up to 40 ha command area, power availability may be considered at maximum of 12 hours.

If the command area is less than 10 ha then 8 hours of power supply can be considered on economic considerations.

Intake Well

This is a civil structure required for guiding the water in the sump well / jacks well. In some cases, this structure is necessary even for small schemes to take care of water level fluctuations in the river. It also provides silt free water for the pumping operations. However, for small lift schemes, this need not be insisted upon.

Suction Pipes for Pump set

The following table may be followed in determining the diameter of suction and delivery pipes for various discharge ranges:

Discharge (lps)	Diameter of Suction pipe (mm)
5	65
10	80
15	100
20	125
30	150
50	200
100	300 (NABARD)

Rising Main

Rising main should be designed based on discharge and total pumping head. It is expected that the total pumping head in normal cases will not exceed 50m. The total length of the rising main should not exceed 3000 m.

The diameter of rising main recommended for RCC pipes and Rigid PVC pipes for various discharges are given below:

(a) RCC Pipes

Discharge (lps)	Dia (mm)	Discharge (lps)*	Dia (mm)
5	100	55	300
10	150	60	300
15	200	65	300
20	200	70	300
25	225	75	300
30	225	80	300
35	250	85	350
40	250	90	350
45	250	95	350
50	250	100	350

*(lps = litre per second)

(b) Rigid PVC Pipes

Discharge (lps)	Dia (mm)	Discharge (lps)	Dia (mm)
5	90	45	225
10	125	50	250
15	150	55	250
20	150	60	250
25	200	65	250
30	200	70	300
35	200	75	300
40	225	80	300

If direct tapping on the rising main is proposed for drawing partial discharge then such tapping should not exceed 2 numbers.

Water hammer control devices in case of large schemes are necessary but in case of small lift irrigation schemes, it may not be required. However, the accessories like air valve, drain valve may be provided as per the topographic conditions.

Apart from the PVC and RCC pipes, HDPE and AC pipes of equivalent class and diameter can also be allowed.

Cast Iron (CI) and Mild Steel (MS) pipes are not recommended except for road crossing and nala crossing.

Delivery Chambers

The delivery chamber of one minute retention capacity can be provided for release of water in the command area.

Distribution System

Generally a chak size of 8 ha may be considered while designing the distribution system.

If scheme economics does not permit use of underground pipes then open channels can be provided. For schemes up to 8 ha. command area only open channels need be provided.

Pumping Machinery

Most of the small lift irrigation schemes require centrifugal pump sets for pumping surface water / groundwater.

In case the fluctuations between low water level and high flood level is less than 4.5 m, then the puppets can be placed in the pump house subject to suitability of site conditions.

If the pump sets are to be operated with shifting arrangements then it should not exceed 20 HP per unit.

The permanent pump house can be provided if the water level in the river is more or less constant throughout the year. This may vary from site to site depending upon the stream hydrology.

The centrifugal pump sets selected should conform to BIS standard i.e., IS:10804-94 for complete pumping system.

The usual formulae that may be applied for calculating the horse power of the centrifugal pump sets and discharge required to be pumped are given as under:

1. Break Horse Power

BHP = (Q x H) / (75 x e)

Where, BHP = Break Horse Power of the Centrifugal Pump set

Q = Discharge in liters per second

H = Total Head in meters (including friction losses)

e = Overall efficiency of Pump set (as percentage)

2. Discharge required to be pumped:

The discharge required for the envisaged command area may be calculated from the following formula: Qr = (28 x A x I) / (R x t)

Where, Qr = Required discharge in litres per second (lps)

A = Crop area in ha

I = Depth of irrigation in cm

R = Rotation period in days

t = Working hours per day

The discharge required to be pumped during different crop seasons can be calculated for kharif, rabi and summer seasons and the peak discharge i.e., maximum discharge should be adopted as the required discharge (Qr).

Number of Stages

From the management point of view, the number of stages should not exceed two. Especially small lift irrigation projects, as far as possible, there should be only single stage pumping.

Project Cost

The project cost of the scheme should be based on the schedule of rates prevailing in the area.

Command Area Map

A map of the command area showing layout drawing of civil works should be attached to the scheme along with a longitudinal section (L Section) for rising main indicating the Low Water Level (LWL), High Flood Level (HFL) position of fixed foundations, pump set, etc.

Small Lift Irrigation Design on Surface Water

A typical design of small lift irrigation scheme on surface water body is given below.

Basic details of the lift irrigation scheme	
Details	Quantity
Command Area (Ha) =	20
Cropping Pattern (ha)	
Kharif:	
Paddy =	10
Makai=	10
Rabi:	
Wheat	8
Gram	5
Makai	2
Summer	
Makai	5
Static Head (m) =	50
Length of rising main (m) =	3000
No of pumping hours (hrs)	12
The required design is as under:	
Max discharge (cum/hr=	52.778
Total Head (m)	69.68
Hp of pump sets =	27.24
No of pump sets=	2
Diameter of rising main (m)	0.143
Class of pipes = varies from 2.5 kg to 8 kg of different lengths.	

The design varies depending on various factors like, command area, static head, length of rising main, cropping pattern etc. The actual design should be based on the field conditions

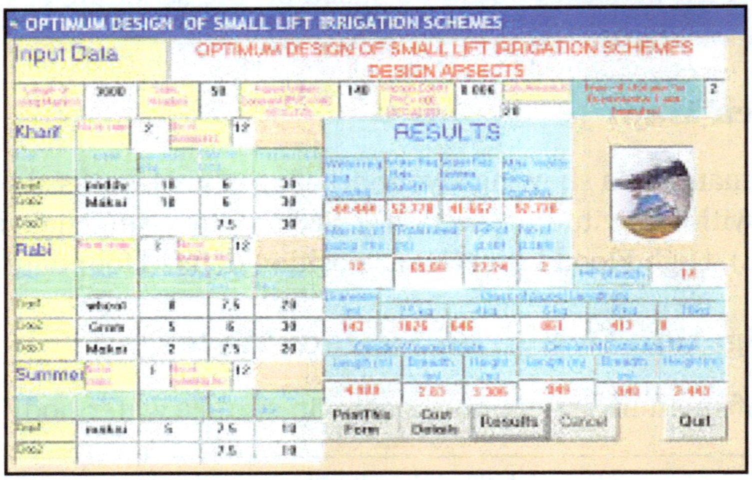

Groundwater Based Small Lift Irrigation Schemes

If the lift irrigation scheme is proposed on the sub surface flow in a given alluvial/ hard rock formation then the intake well has to be designed as per the design criteria adopted for a dug well determined on the basis of aquifer characteristics.

If the source of water is an existing irrigation well, the well discharge and command area to be adopted under small lift irrigation scheme could be similar to one that is as usually considered under a dugwell scheme. For schemes, contemplating well discharge of more than 6 lps in hard rock and 10 lps in alluvial areas, the availability of groundwater from such dug wells has to be ascertained from local groundwater department including stipulations of spacing criteria and clearance of scheme based on categorization of blocks/watersheds etc.

Tube Well Irrigation

Tube well irrigation comes under lift irrigation. Tube well is a small diameter hole drilled.

in the subsoil formation. The cross-sectional area of this type of well is small and unless some mechanical power is used for lifting water, rate of water withdrawal will be low.

The velocity of water withdrawal should be more to maintain high discharge. Obviously the limit of critical velocity is exceeded. Under such conditions the soil particles are dislodged. To prevent clogging of the bore hole a metal pipe with perforations is driven in the hole. It prevents the soil particles from entering the hole. To increase the efficiency of the tube well, in the space between perforated metal tube and soil some straining device is always provided.

The perforations in the metal pipe are necessary only for the depth of water bearing subsoil formations. Other portion of the pipe is generally kept plain. Even though the perforations prevent average soil particles from entering the hole, finer particles enter the hole. To keep the hole clean to avoid settling of particles it is essential to keep the velocity higher (say not less than 1 meter/sec).

The tube well may be taken down to 50 meters in the water bearing stratum. When mechanical power is used to lift water average discharge of tube well is 0.04 m3/sec. The factors which affect the location of tube well are more or less, the same as those discussed for open wells.

Tube well irrigation is very suitable where the subsoil formation is suitable for storing water. Water bearing layer may be of unlimited extent (unconfined) or it may be alternate to some type of impervious layers or may be confined between two impervious layers. In Haryana, Uttar Pradesh, Punjab and Bihar States tube well irrigation is extensively in practice. The obvious reason is that most of the area is made up of alluvium which holds the water to a great extent and there are massive groundwater reservoirs.

Types of Wells

The utilization of groundwater through dug well irrigation is an indigenous form of irrigation. A dug well is a shallow well, with its bottom on a fair depth below the water table, so that water from the surrounding aquifer accumulates in the well. Water collected in the well is lifted to ground surface through a water lift. The masonry lined dug well usually yields 7 to 8 m3 per hour (@ 2 1ps) when operated with a Persian wheel, which is the case for about 20% of the masonry wells. In the remaining dug wells, the water is lifted by animal power with leather or metal buckets, usually bullocks. These wells have very limited discharge rates and this practice is almost dispensed with due to high cost of labour and drudgery.

- Shallow tube wells are drilled to penetrate a shallow aquifer and are usually less than 30 m deep. This depth is only possible when the tube well is placed at the bottom of a dug well, so that it is a dug-cum-tube well. Shallow tube wells are usually equipped with a small centrifugal pump. The filter point wells if Cauvery deltas of Tamil Nadu in Trichy, Thanjavur, Thiruvarur, Nagapatinam districts are shallow tube wells irrigating individual farms. The electric or diesel motor is directly connected to the pump by a belt drive. The centrifugal pump is placed at the surface level and operates mainly in suction mode. These wells usually have a capacity of 20 to 30 m3h-1 (@ 7 lps).

- Medium tube wells are small diameter submersible tube wells equipped with a strainer section. These wells are usually about 45 m in depth although they may be deeper depending on the depth of the aquifer and the capacity desired. They usually have capacities of about 30 to 40 m3h-1 (@ 10 1ps) and are equipped with centrifugal pumps. Water distribution from these wells is through small unlined channels with the following lengths: masonry wells - 30 m; shallow tube wells - 200 m; and medium tube wells - 400 m.
- Deep tube wells have a large diameter and vary in depth from 40 to 300 m. Pumps are sunk into the well, operate in force mode, and are driven by submersible electric engines or by shafts connected to engines at the surface. Deep tube wells have a large discharge capacity varying from 150 to 300 m3h-1 (40 to 80 lps). As discharge capacity increases, the length of the water distribution channels increases accordingly. For example a command area of 100 ha is served by water distribution system of 4 km in state tube well commands of UP, water is distributed through unlined earthen channels.

State Tube Well (STW) Programs

A major instrument of public policy – State Tube well (STW) programs – is devised originally to stimulate groundwater irrigation and to ensure that the access to this communal resource is diffused and is not monopolized by the rural elite. For hard rock regions, open dug wells are technically found to be ideally suited. Average command area of dug wells being rather low (less than 2 hectare), it would mean, in operational terms, government coping with an innumerable number of open dug wells. The other major problem with public tube well programs is their management, efficiency and quality of irrigation service they are able to provide. Numerous field studies, have pointed out poor maintenance, lack of accountability of the tube well operator of the community, domination by local elite, frequent power cuts, delays in repair and procurement of spare-parts, local feuds regarding the right of passage, etc., are amongst the several problems that STW programs suffer from. It was found that small farmers did benefit from public tube wells through improvement in crop pattern, crop yields, and cropping intensity. However, overall experience with public tube wells in various regions of the country is quite disappointing from the point of efficiency. As far as equitable distribution is concerned, the rural elite usually succeed in appropriating most of the benefits of public supply.

Merits of Well and Tube Well Irrigation

1. Well is simplest and cheapest source of irrigation and the poor Indian farmer can easily afford it.

2 Well is an independent source of irrigation and can be used as and when the necessity arises. Canal irrigation, on the other hand, is controlled by other agencies and cannot be used at will.

3. Excessive irrigation by canal leads to the problem of reh which is not the case with well irrigation.

4. There is a limit to the extent of canal irrigation beyond the tail end of the canal while a well can be dug at any convenient place.

5 Several chemicals such as nitrate, chloride, sulphate, etc. are generally found mixed in well water. They add to the fertility of soil when they reach the agricultural field along with well water.

6. The farmer has to pay regularly for canal irrigation which is not the case with well irrigation.

Demerits of Well and Tube Well Irrigation

1. Only limited area can be irrigated. Normally, a well can irrigate 1 to 8 hectares of land.
2. The well may dry up and may be rendered useless for irrigation if excessive water is taken out.
3. In the event of a drought, the ground water level falls and enough water is not available in the well when it is needed the most.
4. Tubewells can draw a lot of groundwater from its neighbouring areas and make the ground dry and unfit for agriculture.
5. Well and tube well irrigation is not possible in areas of brackish groundwater.

Sewage Irrigation

Sewage water is used as potential source irrigation for raising vegetables and fodder crops around the sewage treatment sites, which are directly or indirectly consumed by human beings. Raw Sewage is a rich source of organic and inorganic nutrients for plant growth; sewage farming is quite common in all urban areas in India. Some cities sewage water where industrial effluent is discharged along with may contain toxic metals in high amounts. Thus the composition of domestic sewage may vary with the type of industrial discharge their waste. The analysis of eight chemical parameters of untreated sewage water was carried out in Vidyaranyapuram STP, Mysore, India. Factor analysis was applied on untreated sewage water. Data matrix, pollution factor was found to be the most contributing factor and it showed 22.31% of the total variance (Chloride, BOD, COD and TDS). The second most contributing factor was found to be the nitrification factor, which explained 21.11% of the total variance (pH and Nitrate), whereas the salinization factor contributed 16.98% of the total variance (TS and TSS). Sewage has high values of temperature, pH, hardness, alkalinity, chemical oxygen demand, total soluble salts, nitrates, nitrites and cations like sodium, potassium, calcium and magnesium.

The sewage water has a high fertility load; it adds available N, P, K, Fe, Mn, Zn and Cu to soil, this indicating their significant addition through sewage suggesting use of sewage water as a low grade cheap fertilizer in agriculture. It can drastically reduce the cost due to substitution of chemical fertilizers. The concentration of heavy metals such as Cu, Pb and Co in plant tissue was low compared to limits standards. These heavy metal concentrations are well below hazardous levels. The effect of continuous irrigation with sewage water increases exchangeable cations to a large extent. Sewage water application increases the soil salinity, organic carbon, N, K, Ca, Mg cations to a lot. Soil is a biofilter that can reduce a large part of domestic waste water pollutants, but this filtering increase EC, SAR, Na, Ca and Mg of soil. In addition to these, sewage water also contains significant amounts of toxic metals such as arsenic, chromium, cadmium, copper, lead, nickel, zinc, cobalt, magnesium and iron.

Sewage Farm

Sewage Farm is a plot of land used for the natural biological decontamination of sewage liquids and for raising agricultural crops. In the USSR a distinction is made between municipal and agricultural farms. The former are established on land managed by municipal agencies. They basically perform a sanitary function and differ little in layout from leach fields. Agricultural sewage farms are set up on kolkhozes and sovkhozes for raising crops that use the nutrients contained in sewage liquids. Some are used only in the summer, and others operate year-round.

The sewage farms followed a dewatering concept by J. Hobrecht. In 1869 the Berlin administration made him director of the Berlin Latrine System. Hobrecht divided the city into 12 districts, called radial systems. Each radial system had a pumping station. Pumping stations received domestic, commercial and industrial waste waters as well as precipitation water through gravity flow pipelines. Sewage effluents were conducted from the pumping station through pressure pipelines to sewage farms located outside the city. Some sewage farms were additionally supplied by direct pipelines.

Pressure pipelines discharge waste water at the sewage farms. Waste water is first collected in sedimentation basins made of concrete or earth. Water flows through the tank and most sediments settle to the bottom. Immersion panels hold back floating matter. Sediments settling in the sedimentation basin are regularly evacuated and dewatered at special sludge drying areas. Dewatered sludge was used as a soil conditioner for agriculture and horticulture in early years. The sewage farm trench system is also regularly cleaned, whereby removed sediments are usually deposited directly alongside the trench. After sewage water has passed through the sedimentation basin, e.g. has been mechanically cleaned, it flows through gravity feeders to the terraces.

The natural ground form was not automatically suited for processing sewage waters. Terraces were constructed horizontally or sloping, depending on the surface. They were about 0.25 ha large, and surrounded by embankments. There are three methods of sewage farm treatment. Horizontal terraces are flooded by surrounding distribution ditches. For slope terraces, sewage water overflows the upper bank and irrigates the sloped terrace. Bed terraces with ditch irrigation were also initially used. Waste water flowed through bed terraces in connected parallel furrows, about a meter apart. Only plant roots received water.

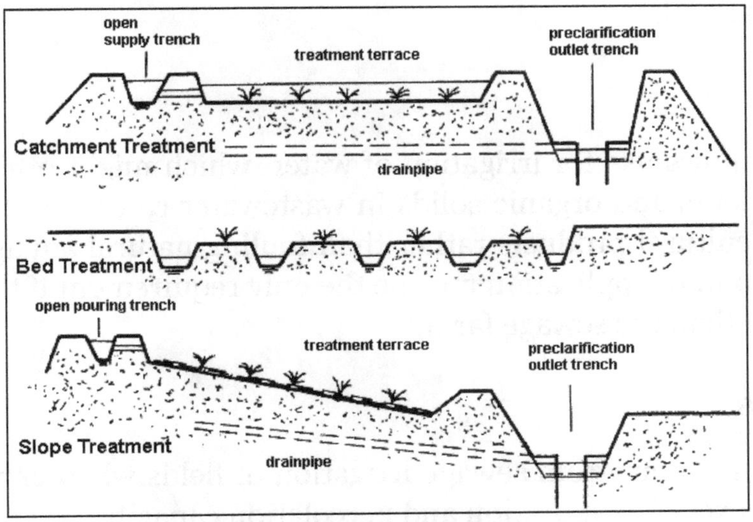

Types of sewage farms.

Wild sewage areas are often found near treatment terraces. The overloading of prepared surfaces can be met by directly diverting unpurified water through sluices onto natural land. Sewage water contents are retained during the passage through the soil, adsorbed in topsoil without humus, and handled chemically and biologically. This process supplies agriculturally useful nutrients. Initial yields were high and the majority of fields were used agriculturally and served their own sewage treatment plots. There was a mixed use of grasslands and field cultivation.

Most sewage farms were provided during construction with drainage pipelines at regular intervals for a faster discharge of filtered and purified water, and to provide for aeration and regeneration of soils as well. Drainage water passes through collecting drains and dewatering trenches into the preclarification outlet trenches. Some water from soil passage percolates into ground water.

Fields are flooded in normal operation, and then left until water seeps away and the soil is re-aerated. The next flooding is begun only after re-aeration is completed. These sewage farm rhythms are also oriented to the growth periods of agricultural crops. Four to eight field treatment cycles a year are possible on grasslands, with 2,000-4,000 mm of sewage water. Areas used for cultivation of winter wheat can only be used once a year, with 100-500 mm of waste water.

Sewage farms were overtaxed with increasing amounts of waste water, intensification of agricultural production, and the closure of other sewage farms. This stimulated some sewage farm operators to establish "intensive filter areas". These are permanently flooded and surrounded by high embankments. An inadequate degree of purification is performed here because aerobic processes cannot take place. These areas were not used agriculturally.

Sewage farm structures were often leveled after sewage treatment use was discontinued. Trenches and terraces were filled with material from the embankments, themselves land-fill material.

Advantage

Sewage farming allows use for irrigation of water, which might otherwise be wasted. Some of the nutrients and organic solids in wastewater can be usefully incorporated into soil and agricultural products rather than fouling natural aquatic environments. Pumping to the point of application may be the only requirement if the village is not at a higher elevation than the sewage farm.

Disadvantages

Polluted runoff may occur from sewage irrigation of fields when entering wastewater and precipitation exceed evaporation and percolation capacity.

Sewage is usually generated at a relatively constant rate, but irrigation is required only during dry weather, and is useful only while temperatures are high enough to promote plant growth. Over-irrigation causes soils to become septic, sour, or *sewage-sick*. Arid climates may allow temporary storage of sewage in holding ponds while the soils dry out during non-growing seasons, but such storage may cause odor and aquatic insect problems, including mosquitoes.

It may be impractical to protect the crops being grown from sewage contact. Even optimum situations like irrigating fruit trees with flow in surface ditches may involve some risk of pathogen transfer from the sewage to the edible fruit by birds, insects, and similar vectors. Pathogen transfer is more likely with ground crops, and practically unavoidable with spray irrigation.

Similar Wastewater Systems

Subsurface distribution piping is problematic since it is vulnerable to root blockage and to damage during soil cultivation. Also obstruction of distribution piping by sewage solids discourages sewage farming when wastewater is not pre-treated as it is typically the case in a septic drain field.

Sewage farming should not be confused with wastewater disposal through infiltration basins or subsurface drains.

Plows or harrows may be used to periodically break up vegetation mats which are slowing surface disposal.

Subsurface disposal typically uses pipes placed deep enough to minimize root penetration and often manages overlying vegetation to avoid growth of plants with deep root systems.

Supplemental Irrigation

Supplemental irrigation (SI) can be defined as the addition to essentially rain-fed crops of small amounts of water during times when rainfall fails to pro-vide sufficient moisture for normal plant growth, in order to improve and stabilize yields. Accordingly, the concept of SI in areas with limited water resources is based on the following three aspects.

Surface water sources are the cheapest supply for supplemental irrigation.

First, water is applied to a rain-fed crop, that would normally produce some yield without irrigation.

Second, since precipitation is the principal source of moisture for rain-fed crops, SI is only applied when precipitation fails to provide essential moisture for improved and stabilized production.

Third, the amount and timing of SI are scheduled not to provide moisture stress free conditions throughout the growing season, but to insure that there is a minimum amount of water available during the critical stages of crop growth that would permit optimal instead of maximum yield.

In many areas, groundwater is being over-exploited and sustainability threatened.

The management of supplemental irrigation is seen as a reverse of that of full or conventional irrigation (FI). In the latter, the principal source of moisture is the fully controlled irrigation water, while the highly variable limited precipitation is only supplementary. Unlike FI, the management of SI is dependent on the precipitation as a basic source of water for the crop.

Water-harvesting is a sustainable source of supple-mental irrigation.

Water resources for supplemental irrigation are mainly surface, but shallow groundwater aquifers are being increasingly used. Of the non-conventional water resources, such as treated sewage effluent, which have potential for the future, water harvesting is an important one.

Improving Production with Supplemental Irrigation

Research results from ICARDA and other institutions in the dry areas, as well as data from farmers' fields, show substantial increases in crop yields in response to the application of relatively small amounts of SI. This increase is possible in both low and high rainfall conditions. Average increases in wheat grain yield under low, medium and high annual rainfall at Tel Hadya were about 400%, 150% and 30% using SI amounts of about 180, 125 and 75 mm, respectively. Generally, optimal SI ranges from 75 mm in areas with annual rainfall of 500 mm, to 250 mm in areas receiving 250 mm rain. Determining the optimal amount of SI under various conditions is discussed later.

When rainfall is low, more water is needed, but the response is greater; however, increases in yield are remark-able even when rainfall is as high as 500mm. The response is greater when rain-fall distribution over the season is poor. However, in all rain-fed areas of the WANA region, some time in the spring there is usually a period of stress, which threatens yield levels. Soil-moisture stress usually starts in March, April or May if total seasonal rainfall is low, average or high, respectively.

Yield is much higher with supplemental irrigation.

In Syria, average wheat yields under rain-fed conditions are only 1.25 t/ha and this is one of the highest in the region. With SI, the average grain yield rises to 3 t/ha. In 1996, over 40% of rain-fed areas were under SI and over half of the 4 million tonnes of national production was attributed to this practice. Supplemental irrigation not onlyincreases yield, but also stabilizes production. The coefficient of variation in production in Syria was reduced from100% in rain-fed crop to 10% when SI was practiced. This stability is of great importance since it secures farmers' income.

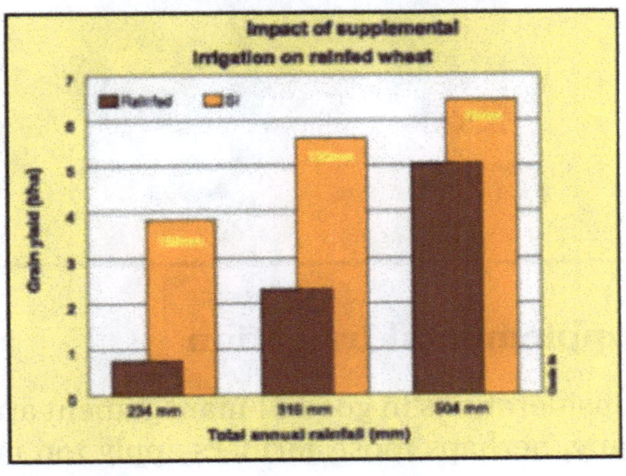

High Water-use Efficiency

Water-use Efficiency (WUE) is a measure of the productivity of the water consumed by the crop. In areas with limited water resources, where water is the greatest limitation to production, WUE is the main criterion for evaluating the performance of agricultural production systems. No longer is productivity per unit area the main objective, since land is not as limiting to production as is water.

Average WUE of rain in producing wheat in the dry areas of WANA is about 0.35 kg grain/m3, although with good management and favorable rainfall(amount and distribution), this can be increased to 1 kg grain/m3. However, water used in SI can be much more efficient. Research at ICARDA showed that a cubic meter of water applied at the right time (when the crop is suffering from moisture stress), combined with good management, could produce more than 2.5 kg of grain over the rain-fed production. This extremely high WUE is mainly attributed to the effectiveness of a small amount of water in alleviating severe moisture stress during the most sensitive stage of crop growth and seed filling. When SI water is applied before such conditions occur, the plant may reach its high yield potential.

In comparison to the productivity of water in fully irrigated areas (when rain-fall effect is negligible), the productivity is higher with SI. In fully irrigated areas with good management, wheat grain yield is about 6 t/ha using 800 mm of water. Thus, the WUE is about 0.75kg/m3, one-third of that under SI with similar management. This suggests that water resources may be better allocated to SI when other physical and economic conditions are favorable.

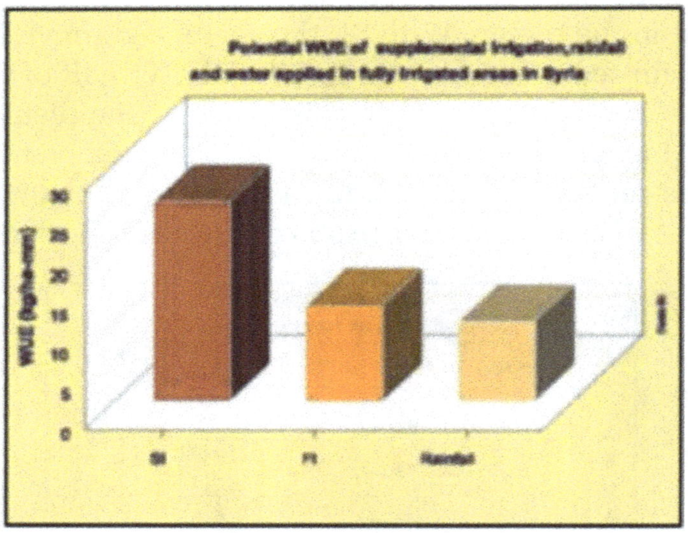

Management of Supplemental Irrigation

The most important considerations in good SI management are when and how much water to apply. Yet many, perhaps most, farmers apply too much water if they can get it. Evidence of the over-use of irrigation water is clear in many dry-area situations, and SI is no exception. Farmers tend to overuse water in SI because of the low water and irrigation costs. The aim of any management pro-gram for SI should be to provide sufficient water to crops at the right time and also to discourage farmers from over-irrigating.

When to use Supplemental Irrigation

Unlike in full irrigation, the time of SI irrigation cannot be determined in advance. This is due to the fact that the basic source of water to rain-fed crops is rainfall, which being variable in amount and distribution, is difficult to predict. Since SI water is best given when soil moisture drops to a critical level, the time of irrigation can be best determined by measuring soil moisture on a regular basis. Unfortunately, there is no simple device that an average farmer can use for this purpose. The well-known tensiometers are not suitable, since SI allows lower soil-moisture potential than the tension meter is able to read correctly; other more sophisticated methods are not appropriate either.

Instead, most farmers in the region rely on personal experience related to the amount of rainfall received and crop appearance. Generally, they tend to irrigate earlier than necessary with more frequent irrigation than needed when they have on-demand water supply.

Cultural Practices

Soil Fertility

SI alone, although it alleviates moisture stress, cannot ensure good performance of the rain-fed agricultural sys-tem. It has to be combined with other farm management practices. Of most importance is the soil fertility; particularly in the Mediterranean region nitro-gen is usually the main deficiency. Absence of nutrient deficiency greatly improves yield and water-use efficiency.

Nitrogen deficiency affects both yield and water-use efficiency.

Under rain-fed conditions, the rate of fertilizer nitrogen needed is not high, and with some water stress, high rates may actually be harmful. Under the Syrian rain-fed conditions, 50 kg N/ha was sufficient. However, with more water applied, the crop responds to nitrogen up to 100 kg N/ha, after which no benefit is obtained. This rate of N greatly improves WUE. It is also important that there is adequate avail-able phosphorus in the soil so that response to N and applied irrigation is not constrained. Other areas may have deficiencies in other elements. It is always important to eliminate these deficiencies to increase yield and WUE.

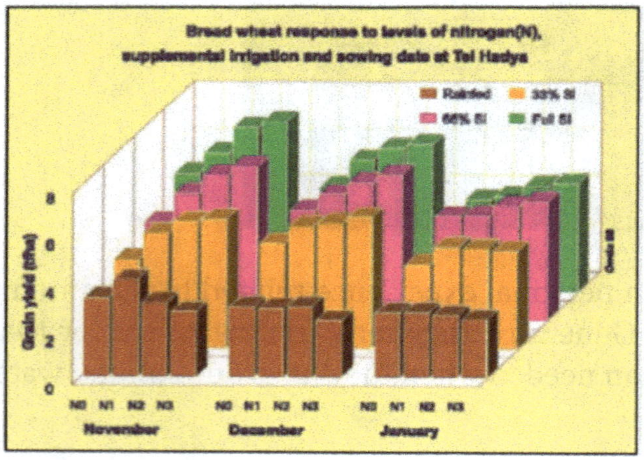

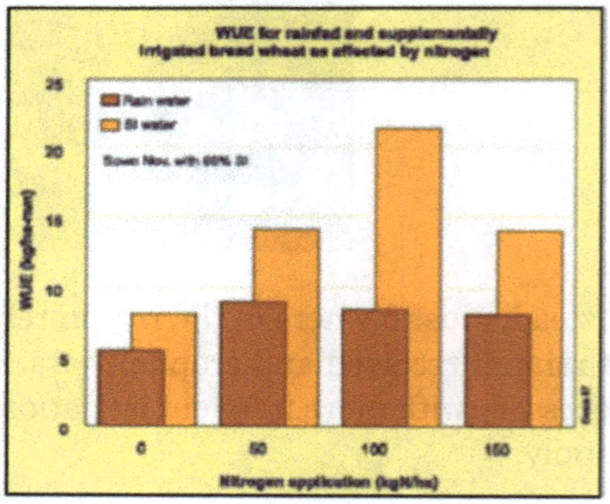

Crop Varieties

Selection of the appropriate crop variety makes a difference under both rain-fed and SI conditions. In rain-fed areas, breeding seeks to produce drought resistant varieties. These perform well under rain-fed conditions, but because they were not developed for SI conditions their response to more reliable water supplies may not be high. A suitable variety for SI is one with good response to limited water application while maintaining some drought resistance.

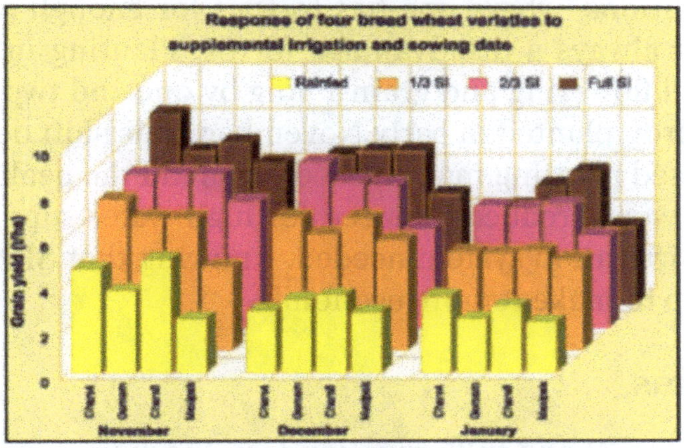

Planting Date

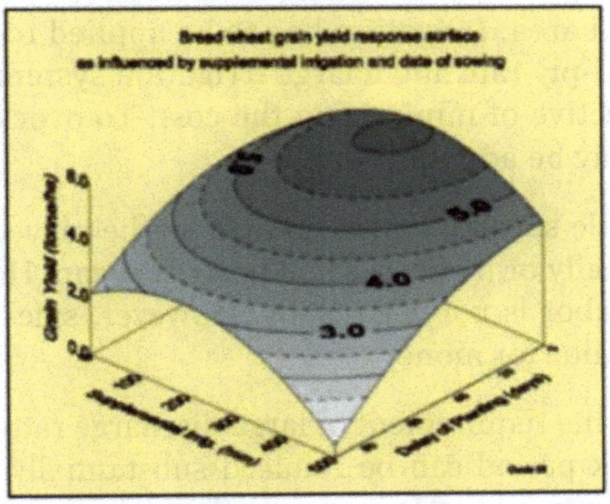

Under rain-fed farming, the earliest planting is usually in November, when enough rain for germination has fallen. Around 15 November is the optimal date for achieving highest yield under rain-fed conditions in the eastern Mediterranean region. Delaying planting after this date has negative impact on yield. Under SI, early planting (1 November) has an advantage in both yield and WUE. With recommended levels of SI and N, wheat yield was decreased substantially by delaying sowing from December to January; for WUE the decrease occurred mainly with the delay from November to December with little decrease afterward. With SI, early planting is possible unlike under rain-fed conditions, where one has to wait for enough rain. Delaying planting, however, is not always a disadvantage for SI. Planting in mid-December and mid-January delayed flowering and grain-filling by one and two weeks, respectively, compared with the crop planted in early November. This shift delays the need for SI, and a system of phased planting may be used to spread the peak water requirements during spring. This would reduce the need for high water supply discharge and reduce the size of the irrigation system needed. Knowing that SI cost is a major issue, this concept can help to make it more economic.

Irrigation Systems

The difference between SI and full irrigation is in the management, not in the physical system. Irrigation systems that are suitable for full irrigation are also technically suitable for SI. There are also economic considerations. Systems for SI are only occasionally used. In most of the rain-fed areas only one to three irrigations may be needed each year. Using a system occasionally rarely justifies large investment. This may be why most SI systems are of the surface type, although they have lower application efficiency and uniformity. More costly systems, such as sprinklers and trickle are mainly acquired for full irrigation in summer but utilized for SI in winter. When labor and water are costly and field is not suitable for surface system, or no full irrigation systems exist, then sprinkler systems are generally used. The objective then is to minimize the cost by appropriate selection of the type and size of the system. One problem can be that the need for water occurs at the same time for all the planted area. Irrigation has to be applied to a large area in a short time. A large water sup-ply rate and a large irrigation system are then needed. This conflicts with the objective of minimizing the cost. To over come this problem the following strategies may be adopted.

1. The use of mobile sprinkler systems that can be moved easily within the field either mechanically or by hand when labor is cheap. Hand-moved systems are suitable when labor is not very costly. However, side-roll and gun sprinklers also give good value for money.

2. The problem of the requirement of large discharge rate and system size during the narrow peak period can be reduced substantially by spreading the peak water requirements over a longer period in late spring. This can be done by using different planting dates in different parts of the field, ranging from early November to late January, and selecting different crops and varieties that require water at different times. Although this may affect the yield and production a little, it will cut the cost of irrigation.

3. During the peak irrigation period, the amount of water required may be divided into two or more applications. A smaller amount is given, but covering the field faster avoids severe stress.

4. Water can be stored in the soil profile before the beginning of the irrigation time. During the early spring, crop consumption rate increases and some soil storage space is created before the critical level is reached. During this time farmers can start irrigating early and fill part of this space, which in turn will delay the time by which next irrigation will be needed. This will educe the peak water need. In fact, some farmers who have small irrigation systems (one hand-moved system) already practice this.

5. When the farm has irrigated summer crops, the system must be designed to suit both full and supplemental irrigation.

6. When the system is only for SI, size and cost may be reduced by considering the optimal water scheduling needs, not the crop's full water requirement. Providing 50% of the full irrigation requirements not only increases WUE and net return, it also cuts water need, irrigation system size and cost.

Hand-moved sprinkler system is low in cost butrequires extensive labor.

Attempts to adapt furrow irrigation for supplementalirrigation of wheat may help in reducing irrigationcost while maintaining high efficiency.

Surface irrigation is very popular. It is labor-inten-sive but requires no capital investment.

References

- David Williams (1991). "The rehabilitation of the River Taff". National Rivers Authority. Retrieved 29, April 2020
- Advantages-disadvantages-main-types-irrigation: lorecentral.org, Retrieved 12, May 2020
- Micro-irrigation: appropedia.org, Retrieved 13, April 2020
- Thomas, Jacqueline M.; Thomas, Torsten; Stuetz, Richard M.; Ashbolt, Nicholas J. (2014). "Your Garden Hose: A Potential Health Risk Due toLegionellaspp. Growth Facilitated by Free-Living Amoebae". Environmental Science & Technology. 48 (17): 10456–10464. doi:10.1021/es502652n. ISSN 0013-936X
- Technical-manual-drip: godavaripipe.com, Retrieved 17, July 2020
- Spring-irrigation, agricultural-engineering: agritech.tnau.ac.in, Retrieved 26, June 2020
- Supplemental-Irrigation-A-Highly-Efficient-Water-Use-Practice-267138098: researchgate.net, Retrieved 21,April 2020
- Wells-and-tube-wells-irrigation-in-india-merit-and-demerits-21109: yourarticlelibrary.com, Retrieved 26,June 2020
- "Welsh History Review - Vol. 14, nos. 1-4 1988-89 Merthyr Tydfil in the mid-Nineteenth Century: the struggle for public health". The National Library of Wales Aberystwyth. Retrieved 23,October 2020
- Soil-and-organics/what-is-gravity-irrigation, growology: gardenculturemagazine.com, Retrieved 10, May 2020
- Bill Lauer (1 January 2004). AWWA Water Operator Field Guide. American Water Works Association. pp. 209–. ISBN 978-1-58321-315-5
- Reed, Sherwood C.; Middlebrooks, E. Joe; Crites, Ronald W. (1988). Natural Systems for Waste Management & Treatment. McGraw-Hill. pp. 220&246. ISBN 0-07-051521-2.

Chapter 5

Irrigation: Water Resources

Water resources are the natural resources of water that are potentially useful to humans. The water present beneath Earth's surface in the spaces of soil is known as groundwater. An aquifer is the saturated zone beneath the water table. This chapter has been carefully written to provide an easy understanding of the various water resources for irrigation.

Water Resources

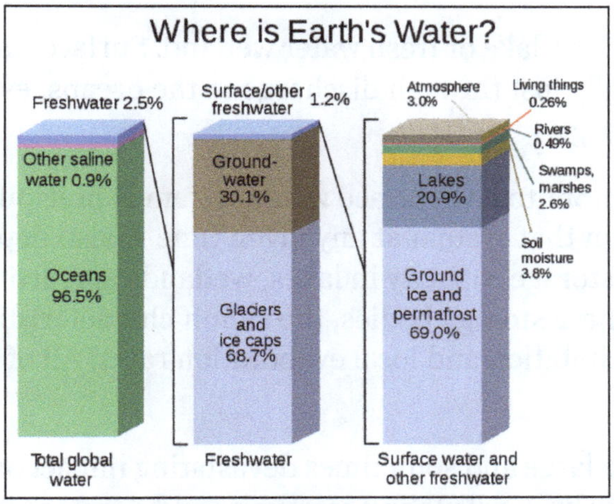

A graphical distribution of the locations of water on Earth. Only 3% of the Earth's water is fresh water. Most of it is in icecaps and glaciers (69%) and groundwater (30%), while all lakes, rivers and swamps combined only account for a small fraction (0.3%) of the Earth's total freshwater reserves.

Water resources are sources of water that are potentially useful. Uses of water include agricultural, industrial, household, recreational and environmental activities. The majority of human uses require fresh water.

97% of the water on the Earth is salt water and only three percent is fresh water; slightly over two thirds of this is frozen in glaciers and polar ice caps. The remaining unfrozen freshwater is found mainly as groundwater, with only a small fraction present above ground or in the air.

Fresh water is a renewable resource, yet the world's supply of groundwater is steadily decreasing, with depletion occurring most prominently in Asia and North America, although it is still unclear how much natural renewal balances this usage, and whether ecosystems are threatened. The framework for allocating water resources to water users (where such a framework exists) is known as water rights.

Sources of Fresh Water

Surface Water

Lake Chungará and Parinacota volcano in northern Chile.

Surface water is water in a river, lake or fresh water wetland. Surface water is naturally replenished by precipitation and naturally lost through discharge to the oceans, evaporation, evapotranspiration and groundwater recharge.

Although the only natural input to any surface water system is precipitation within its watershed, the total quantity of water in that system at any given time is also dependent on many other factors. These factors include storage capacity in lakes, wetlands and artificial reservoirs, the permeability of the soil beneath these storage bodies, the runoff characteristics of the land in the watershed, the timing of the precipitation and local evaporation rates. All of these factors also affect the proportions of water loss.

Human activities can have a large and sometimes devastating impact on these factors. Humans often increase storage capacity by constructing reservoirs and decrease it by draining wetlands. Humans often increase runoff quantities and velocities by paving areas and channelizing stream flow.

The total quantity of water available at any given time is an important consideration. Some human water users have an intermittent need for water. For example, many farms require large quantities of water in the spring, and no water at all in the winter. To supply such a farm with water, a surface water system may require a large storage capacity to collect water throughout the year and release it in a short period of time. Other users have a continuous need for water, such as a power plant that requires water for cooling. To supply such a power plant with water, a surface water system only needs enough storage capacity to fill in when average stream flow is below the power plant's need.

Nevertheless, over the long term the average rate of precipitation within a watershed is the upper bound for average consumption of natural surface water from that watershed.

Natural surface water can be augmented by importing surface water from another watershed through a canal or pipeline. It can also be artificially augmented from any of the other sources listed here, however in practice the quantities are negligible. Humans can also cause surface water to be "lost" (i.e. become unusable) through pollution.

Brazil is the country estimated to have the largest supply of fresh water in the world, followed by Russia and Canada.

Panorama of a natural wetland (Sinclair Wetlands, New Zealand).

Under River Flow

Throughout the course of a river, the total volume of water transported downstream will often be a combination of the visible free water flow together with a substantial contribution flowing through rocks and sediments that underlie the river and its floodplain called the hyporheic zone. For many rivers in large valleys, this unseen component of flow may greatly exceed the visible flow. The hyporheic zone often forms a dynamic interface between surface water and groundwater from aquifers, exchanging flow between rivers and aquifers that may be fully charged or depleted. This is especially significant in karst areas where pot-holes and underground rivers are common.

Groundwater

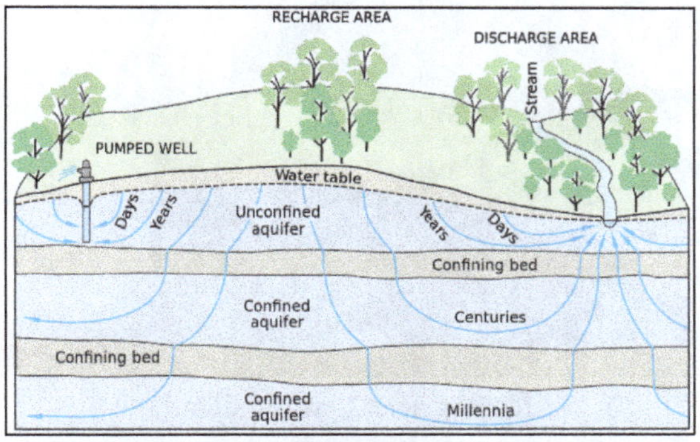

Relative groundwater travel times in the subsurface.

Groundwater is fresh water located in the subsurface pore space of soil and rocks. It is also water that is flowing within aquifers below the water table. Sometimes it is useful to make a distinction between groundwater that is closely associated with surface water and deep groundwater in an aquifer (sometimes called "fossil water").

A shipot is a common water source in Central Ukrainian villages.

Groundwater can be thought of in the same terms as surface water: inputs, outputs and storage.

The critical difference is that due to its slow rate of turnover, groundwater storage is generally much larger (in volume) compared to inputs than it is for surface water. This difference makes it easy for humans to use groundwater unsustainably for a long time without severe consequences. Nevertheless, over the long term the average rate of seepage above a groundwater source is the upper bound for average consumption of water from that source.

The natural input to groundwater is seepage from surface water. The natural outputs from groundwater are springs and seepage to the oceans.

If the surface water source is also subject to substantial evaporation, a groundwater source may become saline. This situation can occur naturally under endorheic bodies of water, or artificially under irrigated farmland. In coastal areas, human use of a groundwater source may cause the direction of seepage to ocean to reverse which can also cause soil salinization. Humans can also cause groundwater to be "lost" (i.e. become unusable) through pollution. Humans can increase the input to a groundwater source by building reservoirs or detention ponds.

Frozen Water

Iceberg near Newfoundland.

Several schemes have been proposed to make use of icebergs as a water source, however to date this has only been done for research purposes. Glacier runoff is considered to be surface water.

The Himalayas, which are often called "The Roof of the World", contain some of the most extensive and rough high altitude areas on Earth as well as the greatest area of glaciers and permafrost outside of the poles. Ten of Asia's largest rivers flow from there, and more than a billion people's livelihoods depend on them. To complicate matters, temperatures there are rising more rapidly than the global average. In Nepal, the temperature has risen by 0.6 degrees Celsius over the last decade, whereas globally, the Earth has warmed approximately 0.7 degrees Celsius over the last hundred years.

Desalination

Desalination is an artificial process by which saline water (generally sea water) is converted to fresh water. The most common desalination processes are distillation and reverse osmosis. Desalination is currently expensive compared to most alternative sources of water, and only a very small fraction of total human use is satisfied by desalination. It is only economically practical for high-valued uses (such as household and industrial uses) in arid areas. The most extensive use is in the Persian Gulf.

Water Uses

Agricultural

It is estimated that 70% of worldwide water is used for irrigation, with 15-35% of irrigation withdrawals being unsustainable. It takes around 2,000 - 3,000 litres of water to produce enough food to satisfy one person's daily dietary need. This is a considerable amount, when compared to that required for drinking, which is between two and five litres. To produce food for the now over 7 billion people who inhabit the planet today requires the water that would fill a canal ten metres deep, 100 metres wide and 2100 kilometres long.

Increasing Water Scarcity

Around fifty years ago, the common perception was that water was an infinite resource. At that time, there were fewer than half the current number of people on the planet. People were not as wealthy as today, consumed fewer calories and ate less meat, so less water was needed to produce their food. They required a third of the volume of water we presently take from rivers. Today, the competition for water resources is much more intense. This is because there are now seven billion people on the planet, their consumption of water-thirsty meat and vegetables is rising, and there is increasing competition for water from industry, urbanisation biofuel crops, and water reliant food items. In the future, even more water will be needed to produce food because the Earth's population is forecast to rise to 9 billion by 2050. An additional 2.5 or 3 billion people, choosing to eat fewer cereals and more meat and vegetables could add an additional five million kilometres to the virtual canal mentioned above.

An assessment of water management in agriculture sector was conducted in 2007 by the International Water Management Institute in Sri Lanka to see if the world had sufficient water to provide food for its growing population. It assessed the current availability of water for agriculture on a global scale and mapped out locations suffering from water scarcity. It found that a fifth of the world's people, more than 1.2 billion, live in areas of physical water scarcity, where there is not enough water to meet all demands. One third of the worlds population does not have access to clean drinking water, which is more than 2.3 billion people. A further 1.6 billion people live in areas experiencing economic water scarcity, where the lack of investment in water or insufficient human capacity make it impossible for authorities to satisfy the demand for water. The report found that it would be possible to produce the food required in future, but that continuation of today's food production and environmental trends would lead to crises in many parts of the world. To avoid a global water crisis, farmers will have to strive to increase productivity to meet growing demands for food, while industry and cities find ways to use water more efficiently.

In some areas of the world, irrigation is necessary to grow any crop at all, in other areas it permits more profitable crops to be grown or enhances crop yield. Various irrigation methods involve different trade-offs between crop yield, water consumption and capital cost of equipment and structures. Irrigation methods such as furrow and overhead sprinkler irrigation are usually less expensive but are also typically less efficient, because much of the water evaporates, runs off or drains below the root zone. Other irrigation methods considered to be more efficient include drip or trickle irrigation, surge irrigation, and some types of sprinkler systems where the sprinklers are operated near ground level. These types of systems, while more expensive, usually offer greater potential to minimize runoff, drainage and evaporation. Any system that is improperly managed can be wasteful, all methods have the potential for high efficiencies under suitable conditions, appropriate irrigation timing and management. Some issues that are often insufficiently considered are salinization of groundwater and contaminant accumulation leading to water quality declines.

As global populations grow, and as demand for food increases in a world with a fixed water supply, there are efforts under way to learn how to produce more food with less water, through improvements in irrigation methods and technologies, agricultural water management, crop types, and water monitoring. Aquaculture is a small but growing agricultural use of water. Freshwater commercial fisheries may also be considered as agricultural uses of water, but have generally been assigned a lower priority than irrigation.

Industrial

A power plant in Poland.

It is estimated that 22% of worldwide water is used in industry. Major industrial users include hydroelectric dams, thermoelectric power plants, which use water for cooling, ore and oil refineries, which use water in chemical processes, and manufacturing plants, which use water as a solvent. Water withdrawal can be very high for certain industries, but consumption is generally much lower than that of agriculture.

Water is used in renewable power generation. Hydroelectric power derives energy from the force of water flowing downhill, driving a turbine connected to a generator. This hydroelectricity is a low-cost, non-polluting, renewable energy source. Significantly, hydroelectric power can also be used for load following unlike most renewable energy sources which are intermittent. Ultimately, the energy in a hydroelectric powerplant is supplied by the sun. Heat from the sun evaporates water, which condenses as rain in higher altitudes and flows downhill. Pumped-storage hydroelectric plants also exist, which use grid electricity to pump water uphill when demand is low, and use the stored water to produce electricity when demand is high.

Hydroelectric power plants generally require the creation of a large artificial lake. Evaporation from this lake is higher than evaporation from a river due to the larger surface area exposed to the elements, resulting in much higher water consumption. The process of driving water through the turbine and tunnels or pipes also briefly removes this water from the natural environment, creating water withdrawal. The impact of this withdrawal on wildlife varies greatly depending on the design of the powerplant.

Pressurized water is used in water blasting and water jet cutters. Also, very high pressure water guns are used for precise cutting. It works very well, is relatively safe, and is not harmful to the environment. It is also used in the cooling of machinery to prevent overheating, or prevent saw blades from overheating. This is generally a very small source of water consumption relative to other uses.

Water is also used in many large scale industrial processes, such as thermoelectric power production, oil refining, fertilizer production and other chemical plant use, and natural gas extraction from shale rock. Discharge of untreated water from industrial uses is pollution. Pollution includes discharged solutes (chemical pollution) and increased water temperature (thermal pollution). Industry requires pure water for many applications and utilizes a variety of purification techniques both in water supply and discharge. Most of this pure water is generated on site, either from natural freshwater or from municipal grey water. Industrial consumption of water is generally much lower than withdrawal, due to laws requiring industrial grey water to be treated and returned to the environment. Thermoelectric powerplants using cooling towers have high consumption, nearly equal to their withdrawal, as most of the withdrawn water is evaporated as part of the cooling process. The withdrawal, however, is lower than in once-through cooling systems.

Domestic

Drinking water.

It is estimated that 8% of worldwide water use is for domestic purposes. These include drinking water, bathing, cooking, toilet flushing, cleaning, laundry and gardening. Basic domestic water requirements have been estimated by Peter Gleick at around 50 liters per person per day, excluding water for gardens. Drinking water is water that is of sufficiently high quality so that it can be consumed or used without risk of immediate or long term harm. Such water is commonly called potable water. In most developed countries, the water supplied to domestic, commerce and industry is all of drinking water standard even though only a very small proportion is actually consumed or used in food preparation.

Recreation

Whitewater rapids.

Sustainable management of water resources (including provision of safe and reliable supplies for drinking water and irrigation, adequate sanitation, protection of aquatic ecosystems, and flood protection) poses enormous challenges in many parts of the world.

Recreational water use is usually a very small but growing percentage of total water use. Recreational water use is mostly tied to reservoirs. If a reservoir is kept fuller than it would otherwise be for recreation, then the water retained could be categorized as recreational usage. Release of water from a few reservoirs is also timed to enhance whitewater boating, which also could be considered a recreational usage. Other examples are anglers, water skiers, nature enthusiasts and swimmers.

Recreational usage is usually non-consumptive. Golf courses are often targeted as using excessive amounts of water, especially in drier regions. It is, however, unclear whether recreational irrigation (which would include private gardens) has a noticeable effect on water resources. This is largely due to the unavailability of reliable data. Additionally, many golf courses utilize either primarily or exclusively treated effluent water, which has little impact on potable water availability.

Some governments, including the Californian Government, have labelled golf course usage as agricultural in order to deflect environmentalists' charges of wasting water. However, using the above figures as a basis, the actual statistical effect of this reassignment is close to zero. In Arizona, an organized lobby has been established in the form of the Golf Industry Association, a group focused on educating the public on how golf impacts the environment.

Recreational usage may reduce the availability of water for other users at specific times and places. For example, water retained in a reservoir to allow boating in the late summer is not available to farmers during the spring planting season. Water released for whitewater rafting may not be available for hydroelectric generation during the time of peak electrical demand.

Environmental

Explicit environment water use is also a very small but growing percentage of total water use. Environmental water may include water stored in impoundments and released for environmental purposes (held environmental water), but more often is water retained in waterways through regulatory limits of abstraction. Environmental water usage includes watering of natural or artificial wetlands, artificial lakes intended to create wildlife habitat, fish ladders, and water releases from reservoirs timed to help fish spawn, or to restore more natural flow regimes.

Like recreational usage, environmental usage is non-consumptive but may reduce the availability of water for other users at specific times and places. For example, water release from a reservoir to help fish spawn may not be available to farms upstream, and water retained in a river to maintain waterway health would not be available to water abstractors downstream.

Water Stress

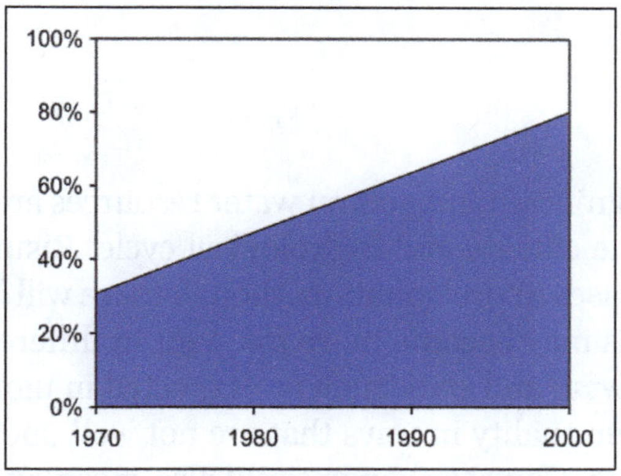

Estimate of the share of people in developing countries with access to drinking water 1970–2000.

The concept of water stress is relatively simple: According to the World Business Council for Sustainable Development, it applies to situations where there is not enough water for all uses, whether agricultural, industrial or domestic. Defining thresholds for stress in terms of available water per capita is more complex, however, entailing assumptions about water use and its efficiency. Nevertheless, it has been proposed that when annual per capita renewable freshwater availability is less than 1,700 cubic meters, countries begin to experience periodic or regular water stress. Below 1,000 cubic meters, water scarcity begins to hamper economic development and human health and well-being.

Population Growth

In 2000, the world population was 6.2 billion. The UN estimates that by 2050 there will be an additional 3.5 billion people with most of the growth in developing countries that already suffer water stress. Thus, water demand will increase unless there are corresponding increases in water conservation and recycling of this vital resource. In building on the data presented here by the UN, the World Bank goes on to explain that access to water for producing food will be one of the main challenges in the decades to come. Access to water will need to be balanced with the importance of managing water itself in a sustainable way while taking into account the impact of climate change, and other environmental and social variables.

Expansion of Business Activity

Business activity ranging from industrialization to services such as tourism and entertainment continues to expand rapidly. This expansion requires increased water services including both supply and sanitation, which can lead to more pressure on water resources and natural ecosystem.

Rapid Urbanization

The trend towards urbanization is accelerating. Small private wells and septic tanks that work well in low-density communities are not feasible within high-density urban areas. Urbanization requires significant investment in water infrastructure in order to deliver water to individuals and to process the concentrations of wastewater – both from individuals and from business. These polluted and contaminated waters must be treated or they pose unacceptable public health risks.

In 60% of European cities with more than 100,000 people, groundwater is being used at a faster rate than it can be replenished. Even if some water remains available, it costs increasingly more to capture it.

Climate Change

Climate change could have significant impacts on water resources around the world because of the close connections between the climate and hydrological cycle. Rising temperatures will increase evaporation and lead to increases in precipitation, though there will be regional variations in rainfall. Both droughts and floods may become more frequent in different regions at different times, and dramatic changes in snowfall and snow melt are expected in mountainous areas. Higher temperatures will also affect water quality in ways that are not well understood. Possible impacts include increased eutrophication. Climate change could also mean an increase in demand for farm irrigation, garden sprinklers, and perhaps even swimming pools. There is now ample evidence that increased hydrologic variability and change in climate has and will continue have a profound impact on the water sector through the hydrologic cycle, water availability, water demand, and water allocation at the global, regional, basin, and local levels.

Depletion of Aquifers

Due to the expanding human population, competition for water is growing such that many of the world's major aquifers are becoming depleted. This is due both for direct human consumption as well as agricultural irrigation by groundwater. Millions of pumps of all sizes are currently extracting groundwater throughout the world. Irrigation in dry areas such as northern China, Nepal and India is supplied by groundwater, and is being extracted at an unsustainable rate. Cities that have experienced aquifer drops between 10 and 50 meters include Mexico City, Bangkok, Beijing, Madras and Shanghai.

Pollution and Water Protection

Water pollution is one of the main concerns of the world today. The governments of numerous countries have striven to find solutions to reduce this problem. Many pollutants threaten water supplies, but the most widespread, especially in developing countries, is the discharge of raw sewage into natural waters; this method of sewage disposal is the most common method in underdeveloped countries, but also is prevalent in quasi-developed countries such as China, India, Nepal and Iran. Sewage, sludge, garbage, and even toxic pollutants are all dumped into the water. Even if sewage is treated, problems still arise. Treated sewage forms sludge, which may be placed in landfills, spread out on land, incinerated or dumped at sea. In addition to sewage, nonpoint source pollution such as agricultural runoff is a significant source of pollution in some parts of the world, along with urban stormwater runoff and chemical wastes dumped by industries and governments.

Polluted water.

Water and Conflicts

Competition for water has widely increased, and it has become more difficult to conciliate the necessities for water supply for human consumption, food production, ecosystems and other uses. Water administration is frequently involved in contradictory and complex problems. Approximately 10% of the worldwide annual runoff is used for human necessities. Several areas of the world are flooded, while others have such low precipitations that human life is almost impossible. As population and development increase, raising water demand, the possibility of problems inside a certain country or region increases, as it happens with others outside the region.

Over the past 25 years, politicians, academics and journalists have frequently predicted that disputes over water would be a source of future wars. Commonly cited quotes include: that of former Egyptian Foreign Minister and former Secretary-General of the United Nations Boutrous Ghali, who forecast, "The next war in the Middle East will be fought over water, not politics"; his successor at the UN, Kofi Annan, who in 2001 said, "Fierce competition for fresh water may well become a source of conflict and wars in the future," and the former Vice President of the World Bank, Ismail Serageldin, who said the wars of the next century will be over water unless significant changes in governance occurred. The water wars hypothesis had its roots in earlier research carried out on a small number of transboundary rivers such as the Indus, Jordan and Nile. These particular rivers became the focus because they had experienced water-related disputes. Specific events cited as evidence include Israel's bombing of Syria's attempts to divert the Jordan's headwaters, and military threats by Egypt against any country building dams in the upstream waters of the Nile. However, while some links made between conflict and water were valid, they did not necessarily represent the norm.

The only known example of an actual inter-state conflict over water took place between 2500 and 2350 BC between the Sumerian states of Lagash and Umma. Water stress has most often led to conflicts at local and regional levels. Tensions arise most often within national borders, in the downstream areas of distressed river basins. Areas such as the lower regions of China's Yellow River or the Chao Phraya River in Thailand, for example, have already been experiencing water stress for several years. Water stress can also exacerbate conflicts and political tensions which are not directly caused by water. Gradual reductions over time in the quality and/or quantity of fresh water can add to the instability of a region by depleting the health of a population, obstructing economic development, and exacerbating larger conflicts.

Shared Water Resources can Promote Collaboration

Water resources that span international boundaries are more likely to be a source of collaboration and cooperation than war. Scientists working at the International Water Management Institute have been investigating the evidence behind water war predictions. Their findings show that, while it is true there has been conflict related to water in a handful of international basins, in the rest of the world's approximately 300 shared basins the record has been largely positive. This is exemplified by the hundreds of treaties in place guiding equitable water use between nations sharing water resources. The institutions created by these agreements can, in fact, be one of the most important factors in ensuring cooperation rather than conflict.

The International Union for the Conservation of Nature (IUCN) published the book *Share: Managing water across boundaries*. One chapter covers the functions of trans-boundary institutions and how they can be designed to promote cooperation, overcome initial disputes and find ways of coping with the uncertainty created by climate change. It also covers how the effectiveness of such institutions can be monitored.

Water Shortages

In 2025, water shortages will be more prevalent among poorer countries where resources are limited and population growth is rapid, such as the Middle East, Africa, and parts of Asia.By 2025, large urban and peri-urban areas will require new infrastructure to provide safe water and adequate sanitation. This suggests growing conflicts with agricultural water users, who currently consume the majority of the water used by humans.

Generally speaking the more developed countries of North America, Europe and Russia will not see a serious threat to water supply by the year 2025, not only because of their relative wealth, but more importantly their populations will be better aligned with available water resources.North Africa, the Middle East, South Africa and northern China will face very severe water shortages due to physical scarcity and a condition of overpopulation relative to their carrying capacity with respect to water supply.Most of South America, Sub-Saharan Africa, Southern China and India will face water supply shortages by 2025; for these latter regions the causes of scarcity will be economic constraints to developing safe drinking water, as well as excessive population growth.

Economic Considerations

Water supply and sanitation require a huge amount of capital investment in infrastructure such as pipe networks, pumping stations and water treatment works. It is estimated that Organisation for Economic Co-operation and Development (OECD) nations need to invest at least USD 200 billion per year to replace aging water infrastructure to guarantee supply, reduce leakage rates and protect water quality.

International attention has focused upon the needs of the developing countries. To meet the Millennium Development Goals targets of halving the proportion of the population lacking access to safe drinking water and basic sanitation by 2015, current annual investment on the order of USD 10 to USD 15 billion would need to be roughly doubled. This does not include investments required for the maintenance of existing infrastructure.

Once infrastructure is in place, operating water supply and sanitation systems entails significant ongoing costs to cover personnel, energy, chemicals, maintenance and other expenses. The sources of money to meet these capital and operational costs are essentially either user fees, public funds or some combination of the two. An increasing dimension to consider is the flexibility of the water supply system.

Groundwater

The entire surface water flow of the Alapaha River near Jennings, Florida going into a sinkhole leading to the Floridan Aquifer groundwater.

Groundwater (or ground water) is the water present beneath Earth's surface in soil pore spaces and in the fractures of rock formations. A unit of rock or an unconsolidated deposit is called an aquifer when it can yield a usable quantity of water. The depth at which soil pore spaces or fractures and voids in rock become completely saturated with water is called the water table. Groundwater is recharged from, and eventually flows to, the surface naturally; natural discharge often occurs at springs and seeps, and can form oases or wetlands. Groundwater is also often withdrawn for agricultural, municipal, and industrial use by constructing and operating extraction wells. The study of the distribution and movement of groundwater is hydrogeology, also called groundwater hydrology.

Typically, groundwater is thought of as water flowing through shallow aquifers, but, in the technical sense, it can also contain soil moisture, permafrost (frozen soil), immobile water in very low permeability bedrock, and deep geothermal or oil formation water. Groundwater is hypothesized to provide lubrication that can possibly influence the movement of faults. It is likely that much of Earth's subsurface contains some water, which may be mixed with other fluids in some instances. Groundwater may not be confined only to Earth. The formation of some of the landforms observed on Mars may have been influenced by groundwater. There is also evidence that liquid water may also exist in the subsurface of Jupiter's moon Europa.

Groundwater is often cheaper, more convenient and less vulnerable to pollution than surface water. Therefore, it is commonly used for public water supplies. For example, groundwater provides the largest source of usable water storage in the United States, and California annually withdraws the largest amount of groundwater of all the states. Underground reservoirs contain far more water than the capacity of all surface reservoirs and lakes in the US, including the Great Lakes. Many municipal water supplies are derived solely from groundwater.

Polluted groundwater is less visible, but more difficult to clean up, than pollution in rivers and lakes. Groundwater pollution most often results from improper disposal of wastes on land. Major sources include industrial and household chemicals and garbage landfills, excessive fertilizers and pesticides used in agriculture, industrial waste lagoons, tailings and process wastewater from mines, industrial fracking, oil field brine pits, leaking underground oil storage tanks and pipelines, sewage sludge and septic systems.

Aquifers

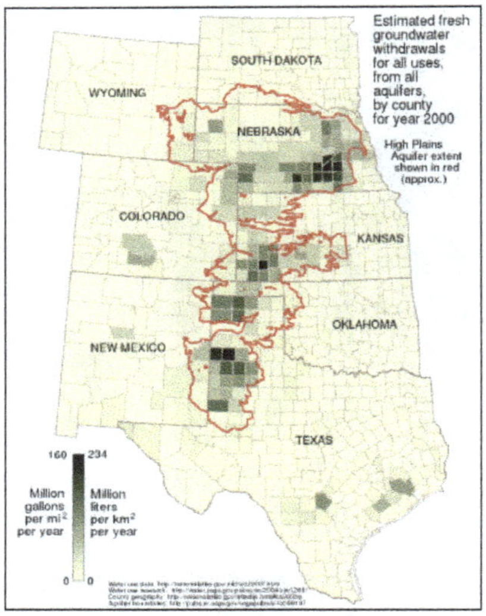

Groundwater withdrawal rates from the Ogallala Aquifer in the Central United States.

An *aquifer* is a layer of porous substrate that contains and transmits groundwater. When water can flow directly between the surface and the saturated zone of an aquifer, the aquifer is unconfined. The deeper parts of unconfined aquifers are usually more saturated since gravity causes water to flow downward.

The upper level of this saturated layer of an unconfined aquifer is called the *water table* or *phreatic surface*. Below the water table, where in general all pore spaces are saturated with water, is the phreatic zone.

Substrate with low porosity that permits limited transmission of groundwater is known as an *aquitard*. An *aquiclude* is a substrate with porosity that is so low it is virtually impermeable to groundwater.

A *confined aquifer* is an aquifer that is overlain by a relatively impermeable layer of rock or substrate such as an aquiclude or aquitard. If a confined aquifer follows a downward grade from its *recharge zone*, groundwater can become pressurized as it flows. This can create artesian wells that flow freely without the need of a pump and rise to a higher elevation than the static water table at the above, unconfined, aquifer.

The characteristics of aquifers vary with the geology and structure of the substrate and topography in which they occur. In general, the more productive aquifers occur in sedimentary geologic formations. By comparison, weathered and fractured crystalline rocks yield smaller quantities of groundwater in many environments. Unconsolidated to poorly cemented alluvial materials that have accumulated as valley-filling sediments in major river valleys and geologically subsiding structural basins are included among the most productive sources of groundwater.

The high specific heat capacity of water and the insulating effect of soil and rock can mitigate the effects of climate and maintain groundwater at a relatively steady temperature. In some places where groundwater temperatures are maintained by this effect at about 10 °C (50 °F), groundwater can be used for controlling the temperature inside structures at the surface. For example, during hot weather relatively cool groundwater can be pumped through radiators in a home and then returned to the ground in another well. During cold seasons, because it is relatively warm, the water can be used in the same way as a source of heat for heat pumps that is much more efficient than using air.

The volume of groundwater in an aquifer can be estimated by measuring water levels in local wells and by examining geologic records from well-drilling to determine the extent, depth and thickness of water-bearing sediments and rocks. Before an investment is made in production wells, test wells may be drilled to measure the depths at which water is encountered and collect samples of soils, rock and water for laboratory analyses. Pumping tests can be performed in test wells to determine flow characteristics of the aquifer.

Water Cycle

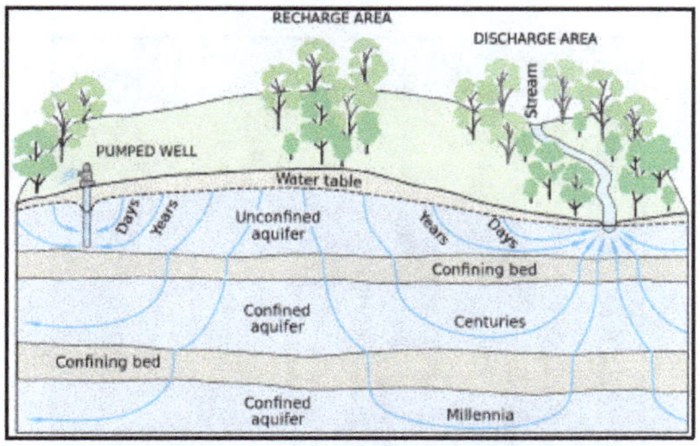

Relative groundwater travel times.

Dzherelo, a common source of drinking water in a Ukrainian village.

Groundwater makes up about twenty percent of the world's fresh water supply, which is about 0.61% of the entire world's water, including oceans and permanent ice. Global groundwater storage is roughly equal to the total amount of freshwater stored in the snow and ice pack, including the north and south poles. This makes it an important resource that can act as a natural storage that can buffer against shortages of surface water, as in during times of drought.

Groundwater is naturally replenished by surface water from precipitation, streams, and rivers when this recharge reaches the water table.

Groundwater can be a long-term 'reservoir' of the natural water cycle (with residence times from days to millennia), as opposed to short-term water reservoirs like the atmosphere and fresh surface water (which have residence times from minutes to years). The figure shows how deep groundwater (which is quite distant from the surface recharge) can take a very long time to complete its natural cycle.

The Great Artesian Basin in central and eastern Australia is one of the largest confined aquifer systems in the world, extending for almost 2 million km^2. By analysing the trace elements in water sourced from deep underground, hydrogeologists have been able to determine that water extracted from these aquifers can be more than 1 million years old.

By comparing the age of groundwater obtained from different parts of the Great Artesian Basin, hydrogeologists have found it increases in age across the basin. Where water recharges the aquifers along the Eastern Divide, ages are young. As groundwater flows westward across the continent, it increases in age, with the oldest groundwater occurring in the western parts. This means that in order to have travelled almost 1000 km from the source of recharge in 1 million years, the groundwater flowing through the Great Artesian Basin travels at an average rate of about 1 metre per year.

Reflective carpet trapping soil water vapor.

Recent research has demonstrated that evaporation of groundwater can play a significant role in the local water cycle, especially in arid regions. Scientists in Saudi Arabia have proposed plans to recapture and recycle this evaporative moisture for crop irrigation. In the opposite photo, a 50-centimeter-square reflective carpet, made of small adjacent plastic cones, was placed in a plant-free dry desert area for five months, without rain or irrigation. It managed to capture and condense enough ground vapor to bring to life naturally buried seeds underneath it, with a green area of about 10% of the carpet area. It is expected that, if seeds were put down before placing this carpet, a much wider area would become green.

Issues

Overview

Certain problems have beset the use of groundwater around the world. Just as river waters have been over-used and polluted in many parts of the world, so too have aquifers. The big difference is that aquifers are out of sight. The other major problem is that water management agencies, when calculating the "sustainable yield" of aquifer and river water, have often counted the same water twice, once in the aquifer, and once in its connected river. This problem, although understood for centuries, has persisted, partly through inertia within government agencies. In Australia, for example, prior to the statutory reforms initiated by the Council of Australian Governments water reform framework in the 1990s, many Australian states managed groundwater and surface water through separate government agencies, an approach beset by rivalry and poor communication.

In general, the time lags inherent in the dynamic response of groundwater to development have been ignored by water management agencies, decades after scientific understanding of the issue was consolidated. In brief, the effects of groundwater overdraft (although undeniably real) may take decades or centuries to manifest themselves. In a classic study in 1982, Bredehoeft and colleagues modeled a situation where groundwater extraction in an intermontane basin withdrew the entire annual recharge, leaving 'nothing' for the natural groundwater-dependent vegetation community. Even when the borefield was situated close to the vegetation, 30% of the original vegetation demand could still be met by the lag inherent in the system after 100 years. By year 500, this had reduced to 0%, signalling complete death of the groundwater-dependent vegetation. The science has been available to make these calculations for decades; however, in general water management agencies have ignored effects that will appear outside the rough timeframe of political elections (3 to 5 years). Marios Sophocleous argued strongly that management agencies must define and use appropriate timeframes in groundwater planning. This will mean calculating groundwater withdrawal permits based on predicted effects decades, sometimes centuries in the future.

As water moves through the landscape, it collects soluble salts, mainly sodium chloride. Where such water enters the atmosphere through evapotranspiration, these salts are left behind. In irrigation districts, poor drainage of soils and surface aquifers can result in water tables' coming to the surface in low-lying areas. Major land degradation problems of soil salinity and waterlogging result, combined with increasing levels of salt in surface waters. As a consequence, major damage has occurred to local economies and environments.

Four important effects are worthy of brief mention. First, flood mitigation schemes, intended to protect infrastructure built on floodplains, have had the unintended consequence of reducing aquifer recharge associated with natural flooding. Second, prolonged depletion of groundwater in extensive aquifers can result in land subsidence, with associated infrastructure damage – as well as, third, saline intrusion. Fourth, draining acid sulphate soils, often found in low-lying coastal plains, can result in acidification and pollution of formerly freshwater and estuarine streams.

Another cause for concern is that groundwater drawdown from over-allocated aquifers has the potential to cause severe damage to both terrestrial and aquatic ecosystems – in some cases very conspicuously but in others quite imperceptibly because of the extended period over which the damage occurs.

Overdraft

Wetlands contrast the arid landscape around Middle Spring, Fish Springs National Wildlife Refuge, Utah.

Groundwater is a highly useful and often abundant resource. However, over-use, or overdraft, can cause major problems to human users and to the environment. The most evident problem (as far as human groundwater use is concerned) is a lowering of the water table beyond the reach of existing wells. As a consequence, wells must be drilled deeper to reach the groundwater; in some places (e.g., California, Texas, and India) the water table has dropped hundreds of feet because of extensive well pumping. In the Punjab region of India, for example, groundwater levels have dropped 10 meters since 1979, and the rate of depletion is accelerating. A lowered water table may, in turn, cause other problems such as groundwater-related subsidence and saltwater intrusion.

Groundwater is also ecologically important. The importance of groundwater to ecosystems is often overlooked, even by freshwater biologists and ecologists. Groundwaters sustain rivers, wetlands, and lakes, as well as subterranean ecosystems within karst or alluvial aquifers.

Not all ecosystems need groundwater, of course. Some terrestrial ecosystems – for example, those of the open deserts and similar arid environments – exist on irregular rainfall and the moisture it delivers to the soil, supplemented by moisture in the air. While there are other terrestrial ecosystems in more hospitable environments where groundwater plays no central role, groundwater is in fact fundamental to many of the world's major ecosystems. Water flows between groundwaters and surface waters. Most rivers, lakes, and wetlands are fed by, and (at other places or times) feed groundwater, to varying degrees. Groundwater feeds soil moisture through percolation, and many terrestrial vegetation communities depend directly on either groundwater or the percolated soil moisture above the aquifer for at least part of each year. Hyporheic zones (the mixing zone of streamwater and groundwater) and riparian zones are examples of ecotones largely or totally dependent on groundwater.

Subsidence

Subsidence occurs when too much water is pumped out from underground, deflating the space below the above-surface, and thus causing the ground to collapse. The result can look like craters on plots of land. This occurs because, in its natural equilibrium state, the hydraulic pressure of groundwater in the pore spaces of the aquifer and the aquitard supports some of the weight of the overlying sediments. When groundwater is removed from aquifers by excessive pumping, pore pressures in the aquifer drop and compression of the aquifer may occur. This compression may be partially recoverable if pressures rebound, but much of it is not. When the aquifer gets compressed, it may cause land subsidence, a drop in the ground surface. The city of New Orleans, Louisiana is actually below sea level today, and its subsidence is partly caused by removal of groundwater from the various aquifer/aquitard systems beneath it. In the first half of the 20th century, the San Joaquin Valley experienced significant subsidence, in some places up to 8.5 metres (28 feet) due to groundwater removal. Cities on river deltas, including Venice in Italy, and Bangkok in Thailand, have experienced surface subsidence; Mexico City, built on a former lake bed, has experienced rates of subsidence of up to 40 cm (1'3") per year.

Seawater Intrusion

In general, in very humid or undeveloped regions, the shape of the water table mimics the slope of the surface. The recharge zone of an aquifer near the seacoast is likely to be inland, often at considerable distance. In these coastal areas, a lowered water table may induce sea water to reverse the flow toward the land. Sea water moving inland is called a saltwater intrusion. In alternative fashion, salt from mineral beds may leach into the groundwater of its own accord.

Pollution

Iron oxide staining caused by reticulation from an unconfined aquifer in karst topography. Perth, Western Australia.

Polluted groundwater is less visible, but more difficult to clean up, than pollution in rivers and lakes. Groundwater pollution most often results from improper disposal of wastes on land. Major sources include industrial and household chemicals and garbage landfills, industrial waste lagoons, tailings and process wastewater from mines, oil field brine pits, leaking underground oil storage tanks and pipelines, sewage sludge and septic systems. Polluted groundwater is mapped by sampling soils and groundwater near suspected or known sources of pollution, to determine the extent of the pollution, and to aid in the design of groundwater remediation systems. Preventing groundwater pollution near potential sources such as landfills requires lining the bottom of a landfill with watertight materials, collecting any leachate with drains, and keeping rainwater off any potential contaminants, along with regular monitoring of nearby groundwater to verify that contaminants have not leaked into the groundwater.

Groundwater pollution, from pollutants released to the ground that can work their way down into groundwater, can create a contaminant plume within an aquifer. Pollution can occur from landfills, naturally occurring arsenic, on-site sanitation systems or other point sources, such as petrol stations or leaking sewers.

Movement of water and dispersion within the aquifer spreads the pollutant over a wider area, its advancing boundary often called a plume edge, which can then intersect with groundwater wells or daylight into surface water such as seeps and springs, making the water supplies unsafe for humans and wildlife. Different mechanism have influence on the transport of pollutants, e.g. diffusion, adsorption, precipitation, decay, in the groundwater. The interaction of groundwater contamination with surface waters is analyzed by use of hydrology transport models.

The danger of pollution of municipal supplies is minimized by locating wells in areas of deep groundwater and impermeable soils, and careful testing and monitoring of the aquifer and nearby potential pollution sources.

Arsenic and Fluoride

Around one-third of the world's population drinks water from groundwater resources. Of this, about 10 percent, approximately 300 million people, obtains water from groundwater resources that are heavily polluted with arsenic or fluoride. These trace elements derive mainly from natural sources by leaching from rock and sediments.

New Method of Identifying Substances that are Hazardous to Health

In 2008, the Swiss Aquatic Research Institute, Eawag, presented a new method by which hazard maps could be produced for geogenic toxic substances in groundwater. This provides an efficient way of determining which wells should be tested.

In 2016, the research group made its knowledge freely available on the Groundwater Assessment Platform GAP. This offers specialists worldwide the possibility of uploading their own measurement data, visually displaying them and producing risk maps for areas of their choice. GAP also serves as a knowledge-sharing forum for enabling further development of methods for removing toxic substances from water.

Government Regulations

In the United States, laws regarding ownership and use of groundwater are generally state laws; however, regulation of groundwater to minimize pollution of groundwater is by both states and the federal-level Environmental Protection Agency. Ownership and use rights to groundwater typically follow one of three main systems:

Rule of Capture

The Rule of Capture provides each landowner the ability to capture as much groundwater as they can put to a beneficial use, but they are not guaranteed any set amount of water. As a result, well-owners are not liable to other landowners for taking water from beneath their land. State laws or regulations will often define "beneficial use", and sometimes place other limits, such as disallowing groundwater extraction which causes subsidence on neighboring property.

Riparian Rights

Limited private ownership rights similar to riparian rights in a surface stream. The amount of groundwater right is based on the size of the surface area where each landowner gets a corresponding amount of the available water. Once adjudicated, the maximum amount of the water right is set, but the right can be decreased if the total amount of available water decreases as is likely during a drought. Landowners may sue others for encroaching upon their groundwater rights, and water pumped for use on the overlying land takes preference over water pumped for use off the land.

Environmental Protection of Groundwater

In November 2006, the Environmental Protection Agency published the groundwater Rule in the United States Federal Register. The EPA was worried that the groundwater system would be vulnerable to contamination from fecal matter. The point of the rule was to keep microbial pathogens out of public water sources. The 2006 groundwater Rule was an amendment of the 1996 Safe Drinking Water Act.

Others

Reasonable Use Rule (American Rule)

This rule does not guarantee the landowner a set amount of water, but allows unlimited extraction as long as the result does not unreasonably damage other wells or the aquifer system. Usually this rule gives great weight to historical uses and prevents new uses that interfere with the prior use.

Groundwater Scrutiny Upon Real Estate Property Transactions in the US

In the US, upon commercial real estate property transactions both groundwater and soil are the subjects of scrutiny, with a Phase I Environmental Site Assessment normally being prepared to investigate and disclose potential pollution issues. In the San Fernando Valley of California, real estate contracts for property transfer below the Santa Susana Field Laboratory (SSFL) and eastward have clauses releasing the seller from liability for groundwater contamination consequences from existing or future pollution of the Valley Aquifer.

Aquifer

An aquifer is an underground layer of water-bearing permeable rock, rock fractures or unconsolidated materials (gravel, sand, or silt) from which groundwater can be extracted using a water well. The study of water flow in aquifers and the characterization of aquifers is called hydrogeology.

Related terms include aquitard, which is a bed of low permeability along an aquifer, and aquiclude (or *aquifuge*), which is a solid, impermeable area underlying or overlying an aquifer. If the impermeable area overlies the aquifer, pressure could cause it to become a confined aquifer.

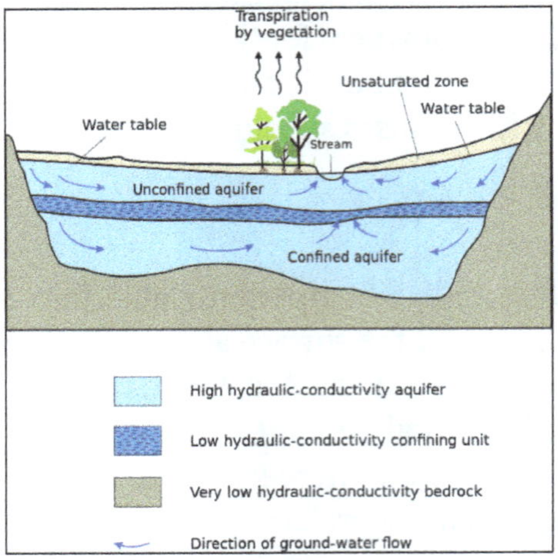

Typical aquifer cross-section.

Depth

Aquifers may occur at various depths. Those closer to the surface are not only more likely to be used for water supply and irrigation, but are also more likely to be topped up by the local rainfall. Many desert areas have limestone hills or mountains within them or close to them that can be exploited as groundwater resources. Part of the Atlas Mountains in North Africa, the Lebanon and Anti-Lebanon ranges between Syria and Lebanon, the Jebel Akhdar (Oman) in Oman, parts of the Sierra Nevada and neighboring ranges in the United States' Southwest, have shallow aquifers that are exploited for their water. Overexploitation can lead to the exceeding of the practical sustained yield; i.e., more water is taken out than can be replenished. Along the coastlines of certain countries, such as Libya and Israel, increased water usage associated with population growth has caused a lowering of the water table and the subsequent contamination of the groundwater with saltwater from the sea.

The beach provides a model to help visualize an aquifer. If a hole is dug into the sand, very wet or saturated sand will be located at a shallow depth. This hole is a crude well, the wet sand represents an aquifer, and the level to which the water rises in this hole represents the water table.

In 2013 large freshwater aquifers were discovered under continental shelves off Australia, China, North America and South Africa. They contain an estimated half a million cubic kilometers of "low salinity" water that could be economically processed into potable water. The reserves formed when ocean levels were lower and rainwater made its way into the ground in land areas that were not submerged until the ice age ended 20,000 years ago. The volume is estimated to be 100x the amount of water extracted from other aquifers since 1900.

Classification

The above diagram indicates typical flow directions in a cross-sectional view of a simple confined or unconfined aquifer system. The system shows two aquifers with one aquitard (a confining or impermeable layer) between them, surrounded by the bedrock *aquiclude*, which is in contact with a gaining stream (typical in humid regions). The water table and unsaturated zone are also illustrated. An *aquitard* is a zone within the earth that restricts the flow of groundwater from one aquifer to another. An aquitard can sometimes, if completely impermeable, be called an *aquiclude* or *aquifuge.* Aquitards are composed of layers of either clay or non-porous rock with low hydraulic conductivity.

Saturated Versus Unsaturated

Groundwater can be found at nearly every point in the Earth's shallow subsurface to some degree, although aquifers do not necessarily contain fresh water. The Earth's crust can be divided into two regions: the *saturated zone* or *phreatic zone* (e.g., aquifers, aquitards, etc.), where all available spaces are filled with water, and the *unsaturated zone* (also called the vadose zone), where there are still pockets of air that contain some water, but can be filled with more water.

Saturated means the pressure head of the water is greater than atmospheric pressure (it has a gauge pressure > 0). The definition of the water table is the surface where the pressure head is equal to atmospheric pressure (where gauge pressure = 0).

Unsaturated conditions occur above the water table where the pressure head is negative (absolute pressure can never be negative, but gauge pressure can) and the water that incompletely fills the pores of the aquifer material is under suction. The water content in the unsaturated zone is held in place by surface adhesive forces and it rises above the water table (the zero-gauge-pressure isobar) by capillary action to saturate a small zone above the phreatic surface (the capillary fringe) at less than atmospheric pressure. This is termed tension saturation and is not the same as saturation on a water-content basis. Water content in a capillary fringe decreases with increasing distance from the phreatic surface. The capillary head depends on soil pore size. In sandy soils with larger pores, the head will be less than in clay soils with very small pores. The normal capillary rise in a clayey soil is less than 1.80 m (six feet) but can range between 0.3 and 10 m (one and 30 ft).

The capillary rise of water in a small-diameter tube involves the same physical process. The water table is the level to which water will rise in a large-diameter pipe (e.g., a well) that goes down into the aquifer and is open to the atmosphere.

Aquifers Versus Aquitards

Aquifers are typically saturated regions of the subsurface that produce an economically feasible quantity of water to a well or spring (e.g., sand and gravel or fractured bedrock often make good aquifer materials).

An aquitard is a zone within the earth that restricts the flow of groundwater from one aquifer to another. A completely impermeable aquitard is called an aquiclude or aquifuge. Aquitards comprise layers of either clay or non-porous rock with low hydraulic conductivity.

In mountainous areas (or near rivers in mountainous areas), the main aquifers are typically unconsolidated alluvium, composed of mostly horizontal layers of materials deposited by water processes (rivers and streams), which in cross-section (looking at a two-dimensional slice of the aquifer) appear to be layers of alternating coarse and fine materials. Coarse materials, because of the high energy needed to move them, tend to be found nearer the source (mountain fronts or rivers), whereas the fine-grained material will make it farther from the source (to the flatter parts of the basin or overbank areas - sometimes called the pressure area). Since there are less fine-grained deposits near the source, this is a place where aquifers are often unconfined (sometimes called the forebay area), or in hydraulic communication with the land surface.

Confined Versus Unconfined

There are two end members in the spectrum of types of aquifers; *confined* and *unconfined* (with semi-confined being in between). Unconfined aquifers are sometimes also called *water table* or *phreatic* aquifers, because their upper boundary is the water table or phreatic surface. Typically (but not always) the shallowest aquifer at a given location is unconfined, meaning it does not have a confining layer (an aquitard or aquiclude) between it and the surface. The term "perched" refers to ground water accumulating above a low-permeability unit or strata, such as a clay layer. This term is generally used to refer to a small local area of ground water that occurs at an elevation higher than a regionally extensive aquifer. The difference between perched and unconfined aquifers is their size (perched is smaller). Confined aquifers are aquifers that are overlain by a confining layer, often made up of clay. The confining layer might offer some protection from surface contamination.

If the distinction between confined and unconfined is not clear geologically (i.e., if it is not known if a clear confining layer exists, or if the geology is more complex, e.g., a fractured bedrock aquifer), the value of storativity returned from an aquifer test can be used to determine it (although aquifer tests in unconfined aquifers should be interpreted differently than confined ones). Confined aquifers have very low storativity values (much less than 0.01, and as little as 10^{-5}), which means that the aquifer is storing water using the mechanisms of aquifer matrix expansion and the compressibility of water, which typically are both quite small quantities. Unconfined aquifers have storativities (typically then called specific yield) greater than 0.01 (1% of bulk volume); they release water from storage by the mechanism of actually draining the pores of the aquifer, releasing relatively large amounts of water (up to the drainable porosity of the aquifer material, or the minimum volumetric water content).

Isotropic Versus Anisotropic

In isotropic aquifers or aquifer layers the hydraulic conductivity (K) is equal for flow in all directions, while in anisotropic conditions it differs, notably in horizontal (Kh) and vertical (Kv) sense.

Semi-confined aquifers with one or more aquitards work as an anisotropic system, even when the separate layers are isotropic, because the compound Kh and Kv values are different.

When calculating flow to drains or flow to wells in an aquifer, the anisotropy is to be taken into account lest the resulting design of the drainage system may be faulty.

Groundwater in Rock Formations

Groundwater may exist in *underground rivers* (e.g., caves where water flows freely underground). This may occur in eroded limestone areas known as karst topography, which make up only a small percentage of Earth's area. More usual is that the pore spaces of rocks in the subsurface are simply saturated with water — like a kitchen sponge — which can be pumped out for agricultural, industrial, or municipal uses.

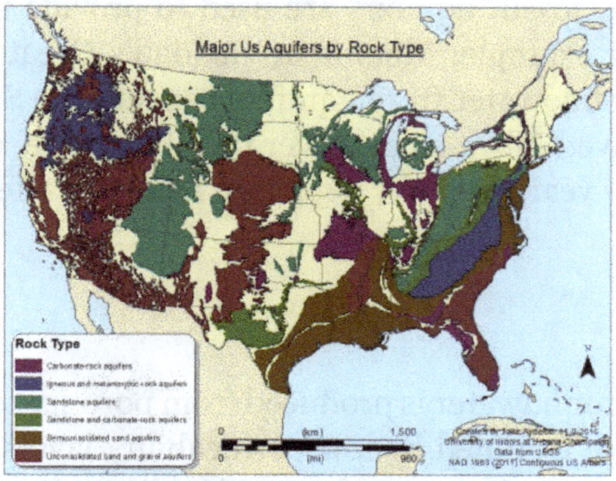

Map of major US aquifers by rock type.

If a rock unit of low porosity is highly fractured, it can also make a good aquifer (via fissure flow), provided the rock has a hydraulic conductivity sufficient to facilitate movement of water. Porosity is important, but, *alone*, it does not determine a rock's ability to act as an aquifer. Areas of the Deccan Traps (a basaltic lava) in west central India are good examples of rock formations with high porosity but low permeability, which makes them poor aquifers. Similarly, the micro-porous (Upper Cretaceous) Chalk of south east England, although having a reasonably high porosity, has a low grain-to-grain permeability, with its good water-yielding characteristics mostly due to micro-fracturing and fissuring.

Human Dependence on Groundwater

Most land areas on Earth have some form of aquifer underlying them, sometimes at significant depths. In some cases, these aquifers are rapidly being depleted by the human population.

Fresh-water aquifers, especially those with limited recharge by snow or rain, also known as meteoric water, can be over-exploited and depending on the local hydrogeology, may draw in non-potable water or saltwater intrusion from hydraulically connected aquifers or surface water bodies. This can be a serious problem, especially in coastal areas and other areas where aquifer pumping is excessive. In some areas, the ground water can become contaminated by arsenic and other mineral poisons.

Aquifers are critically important in human habitation and agriculture. Deep aquifers in arid areas have long been water sources for irrigation (see Ogallala below). Many villages and even large cities draw their water supply from wells in aquifers.

Municipal, irrigation, and industrial water supplies are provided through large wells. Multiple wells for one water supply source are termed "wellfields", which may withdraw water from confined or unconfined aquifers. Using ground water from deep, confined aquifers provides more protection from surface water contamination. Some wells, termed "collector wells," are specifically designed to induce infiltration of surface (usually river) water.

Aquifers that provide sustainable fresh groundwater to urban areas and for agricultural irrigation are typically close to the ground surface (within a couple of hundred metres) and have some recharge by fresh water. This recharge is typically from rivers or meteoric water (precipitation) that percolates into the aquifer through overlying unsaturated materials.

Occasionally, sedimentary or "fossil" aquifers are used to provide irrigation and drinking water to urban areas. In Libya, for example, Muammar Gaddafi's Great Manmade River project has pumped large amounts of groundwater from aquifers beneath the Sahara to populous areas near the coast. Though this has saved Libya money over the alternative, desalination, the aquifers are likely to run dry in 60 to 100 years. Aquifer depletion has been cited as one of the causes of the food price rises of 2011.

Subsidence

In unconsolidated aquifers, groundwater is produced from pore spaces between particles of gravel, sand, and silt. If the aquifer is confined by low-permeability layers, the reduced water pressure in the sand and gravel causes slow drainage of water from the adjoining confining layers. If these confining layers are composed of compressible silt or clay, the loss of water to the aquifer reduces the water pressure in the confining layer, causing it to compress from the weight of overlying geologic materials. In severe cases, this compression can be observed on the ground surface as subsidence. Unfortunately, much of the subsidence from groundwater extraction is permanent (elastic rebound is small). Thus, the subsidence is not only permanent, but the compressed aquifer has a permanently reduced capacity to hold water.

Saltwater Intrusion

Aquifers near the coast have a lens of freshwater near the surface and denser seawater under freshwater. Seawater penetrates the aquifer diffusing in from the ocean and is denser than freshwater. For porous (i.e., sandy) aquifers near the coast, the thickness of freshwater atop saltwater is about 40 feet (12 m) for every 1 ft (0.30 m) of freshwater head above sea level. This relationship is called the Ghyben-Herzberg equation. If too much ground water is pumped near the coast, salt-water may intrude into freshwater aquifers causing contamination of potable freshwater supplies. Many coastal aquifers, such as the Biscayne Aquifer near Miami and the New Jersey Coastal Plain aquifer, have problems with saltwater intrusion as a result of overpumping and sea level rise.

Salination

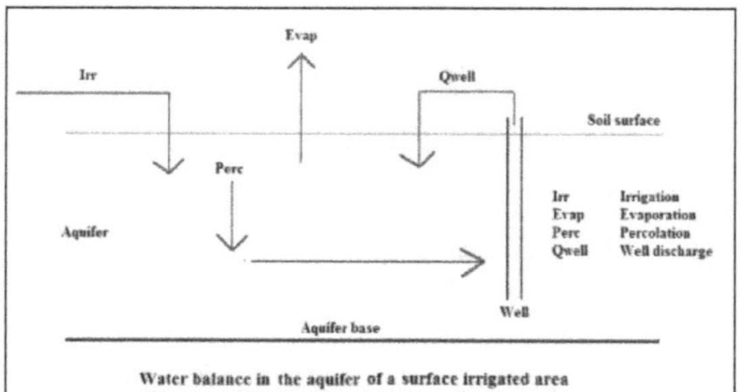

Diagram of a water balance of the aquifer.

Aquifers in surface irrigated areas in semi-arid zones with reuse of the unavoidable irrigation water losses percolating down into the underground by supplemental irrigation from wells run the risk of salination.

Surface irrigation water normally contains salts in the order of 0.5 g/l or more and the annual irrigation requirement is in the order of 10000 m^3/ha or more so the annual import of salt is in the order of 5000 kg/ha or more.

Under the influence of continuous evaporation, the salt concentration of the aquifer water may increase continually and eventually cause an environmental problem.

For salinity control in such a case, annually an amount of drainage water is to be discharged from the aquifer by means of a subsurface drainage system and disposed of through a safe outlet. The drainage system may be *horizontal* (i.e. using pipes, tile drains or ditches) or *vertical* (drainage by wells). To estimate the drainage requirement, the use of a groundwater model with an agro-hydro-salinity component may be instrumental, e.g. SahysMod.

Examples

The Great Artesian Basin situated in Australia is arguably the largest groundwater aquifer in the world (over 1.7 million km^2). It plays a large part in water supplies for Queensland and remote parts of South Australia.

The Guarani Aquifer, located beneath the surface of Argentina, Brazil, Paraguay, and Uruguay, is one of the world's largest aquifer systems and is an important source of fresh water. Named after the Guarani people, it covers 1,200,000 km^2, with a volume of about 40,000 km^3, a thickness of between 50 m and 800 m and a maximum depth of about 1,800 m.

Aquifer depletion is a problem in some areas, and is especially critical in northern Africa; see the Great Manmade River project of Libya for an example. However, new methods of groundwater management such as artificial recharge and injection of surface waters during seasonal wet periods has extended the life of many freshwater aquifers, especially in the United States.

The Ogallala Aquifer of the central United States is one of the world's great aquifers, but in places it is being rapidly depleted by growing municipal use, and continuing agricultural use. This huge aquifer, which underlies portions of eight states, contains primarily fossil water from the time of the last glaciation. Annual recharge, in the more arid parts of the aquifer, is estimated to total only about 10 percent of annual withdrawals. According to a 2013 report by research hydrologist Leonard F. Konikow at the United States Geological Survey (USGC), the depletion between 2001–2008, inclusive, is about 32 percent of the cumulative depletion during the entire 20th century (Konikow 2013:22)." In the United States, the biggest users of water from aquifers include agricultural irrigation and oil and coal extraction. "Cumulative total groundwater depletion in the United States accelerated in the late 1940s and continued at an almost steady linear rate through the end of the century. In addition to widely recognized environmental consequences, groundwater depletion also adversely impacts the long-term sustainability of groundwater supplies to help meet the Nation's water needs."

An example of a significant and sustainable carbonate aquifer is the Edwards Aquifer in central Texas. This carbonate aquifer has historically been providing high quality water for nearly 2 million people, and even today, is full because of tremendous recharge from a number of area streams, rivers and lakes. The primary risk to this resource is human development over the recharge areas.

Discontinuous sand bodies at the base of the McMurray Formation in the Athabasca Oil Sands region of northeastern Alberta, Canada, are commonly referred to as the Basal Water Sand (BWS) aquifers. Saturated with water, they are confined beneath impermeable bitumen-saturated sands that are exploited to recover bitumen for synthetic crude oil production. Where they are deep-lying and recharge occurs from underlying Devonian formations they are saline, and where they are shallow and recharged by meteoric water they are non-saline. The BWS typically pose problems for the recovery of bitumen, whether by open-pit mining or by *in situ* methods such as steam-assisted gravity drainage (SAGD), and in some areas they are targets for waste-water injection.

Groundwater Recharge

Groundwater recharge or deep drainage or deep percolation is a hydrologic process where water moves downward from surface water to groundwater. Recharge is the primary method through which water enters an aquifer. This process usually occurs in the vadose zone below plant roots and is often expressed as a flux to the water table surface. Recharge occurs both naturally (through the water cycle) and through anthropogenic processes (i.e., "artificial groundwater recharge"), where rainwater and or reclaimed water is routed to the subsurface.

Processes

Groundwater is recharged naturally by rain and snow melt and to a smaller extent by surface water (rivers and lakes). Recharge may be impeded somewhat by human activities including paving, development, or logging. These activities can result in loss of topsoil resulting in reduced water infiltration, enhanced surface runoff and reduction in recharge. Use of groundwaters, especially for irrigation, may also lower the water tables. Groundwater recharge is an important process for sustainable groundwater management, since the volume-rate abstracted from an aquifer in the long term should be less than or equal to the volume-rate that is recharged.

Recharge can help move excess salts that accumulate in the root zone to deeper soil layers, or into the groundwater system. Tree roots increase water saturation into groundwater reducing water runoff. Flooding temporarily increases river bed permeability by moving clay soils downstream, and this increases aquifer recharge.

Artificial groundwater recharge is becoming increasingly important in India, where over-pumping of groundwater by farmers has led to underground resources becoming depleted. In 2007, on the recommendations of the International Water Management Institute, the Indian government allocated Rs 1800 crore (US$400million) to fund dug-well recharge projects (a dug-well is a wide, shallow well, often lined with concrete) in 100 districts within seven states where water stored in hard-rock aquifers had been over-exploited. Another environmental issue is the disposal of waste through the water flux such as dairy farms, industrial, and urban runoff.

Wetlands

Wetlands help maintain the level of the water table and exert control on the hydraulic head (O'Brien 1988; Winter 1988). This provides force for groundwater recharge and discharge to other waters as well. The extent of groundwater recharge by a wetland is dependent upon soil, vegetation, site, perimeter to volume ratio, and water table gradient (Carter and Novitzki 1988; Weller 1981).

Groundwater recharge occurs through mineral soils found primarily around the edges of wetlands (Verry and Timmons 1982) The soil under most wetlands is relatively impermeable. A high perimeter to volume ratio, such as in small wetlands, means that the surface area through which water can infiltrate into the groundwater is high (Weller 1981). Groundwater recharge is typical in small wetlands such as prairie potholes, which can contribute significantly to recharge of regional groundwater resources (Weller 1981). Researchers have discovered groundwater recharge of up to 20% of wetland volume per season (Weller 1981).

Estimation Methods

Rates of groundwater recharge are difficult to quantify since other related processes, such as evaporation, transpiration (or evapotranspiration) and infiltration processes must first be measured or estimated to determine the balance.

Physical

Physical methods use the principles of soil physics to estimate recharge. The *direct* physical methods are those that attempt to actually measure the volume of water passing below the root zone. *Indirect* physical methods rely on the measurement or estimation of soil physical parameters, which along with soil physical principles, can be used to estimate the potential or actual recharge. After months without rain the level of the rivers under humid climate is low and represents solely drained groundwater. Thus, the recharge can be calculated from this base flow if the catchment area is known.

Chemical

Chemical methods use the presence of relatively inert water-soluble substances, such as an isotopic tracer or chloride, moving through the soil, as deep drainage occurs.

Numerical Models

Recharge can be estimated using numerical methods, using such codes as Hydrologic Evaluation of Landfill Performance, UNSAT-H, SHAW, WEAP, and MIKE SHE. The 1D-program HYDRUS1D is available online. The codes generally use climate and soil data to arrive at a recharge estimate and use the Richards equation in some form to model groundwater flow in the vadose zone.

Reclaimed Water

Reclaimed water or recycled water, is former wastewater (sewage) that is treated to remove solids and impurities, and used in sustainable landscaping irrigation, to recharge groundwater aquifers, to meet commercial and industrial water needs, and for drinking. The purpose of these processes is water conservation and sustainability, rather than discharging the treated water to surface waters such as rivers and oceans. In some cases, recycled water can be used for streamflow augmentation to benefit ecosystems and improve aesthetics. One example of this is along Calera Creek in the City of Pacifica, CA.

Samples of different types of (waste)water, starting with raw sewage then plant effluent and finally reclaimed water (after several treatment steps).

Definition

Reclaimed waste can be defined as:

"The end product of wastewater reclamation that meets water quality requirements for biodegradable materials, suspended matter and pathogens."

Simply stated, reclaimed water is water that is used more than one time before it passes back into the natural water cycle. Scientifically-proven advances in water technology allow communities to reuse water for many different purposes, including industrial, irrigation, and drinking. The water is treated differently depending upon the source and use of the water and how it gets delivered.

Cycled repeatedly through the planetary hydrosphere, all water on Earth is recycled water, but the terms "recycled water" or "reclaimed water" typically mean wastewater sent from a home or business through a pipeline system to a treatment facility, where it is treated to a level consistent with its intended use. The water is then routed directly to a recycled water system for uses such as irrigation or industrial cooling.

Usage

There are examples of communities that have safely used recycled water for many years. Los Angeles County's sanitation districts have provided treated wastewater for landscape irrigation in parks and golf courses since 1929. The first reclaimed water facility in California was built at San Francisco's Golden Gate Park in 1932. The Water Replenishment District of Southern California was the first groundwater agency to obtain permitted use of recycled water for groundwater recharge in 1962. The Irvine Ranch Water District (IRWD) was the first water district in California to receive an unrestricted use permit from the state for its recycled water; such a permit means that water can be used for any purpose except drinking. IRWD maintains one of the largest recycled water systems in the nation with more than 400 miles serving more than 4,500 metered connections. The Irvine Ranch Water District and Orange County Water District in Southern California are established leaders in recycled water. Further, the Orange County Water District, located in Orange County, and in other locations throughout the world such as Singapore, water is given more advanced treatments and is used indirectly for drinking.

In spite of quite simple methods that incorporate the principles of water-sensitive urban design (WSUD) for easy recovery of stormwater runoff, there remains a common perception that reclaimed water must involve sophisticated and technically complex treatment systems, attempting to recover the most complex and degraded types of sewage. As this effort is driven by sustainability factors, this type of implementation should inherently be associated with point source solutions, where it is most economical to achieve the expected outcomes. Harvesting of stormwater or rainwater can be an extremely simple to comparatively complex, as well as energy and chemical intensive, recovery of more contaminated sewage.

Terminology

Effluent storage tank from where treated effluent (after constructed wetland) is pumped away for irrigation, Haran-Al-Awamied, Syria.

There is no one-size-fits-all solution to water reuse, but there are many safe and scientifically-proven options that allow communities to sustain their local water supplies. Below are terms scientists and water experts use to describe some of these reclaimed water options:

Reused water is water used more than once or recycled.

Potable water is drinking water.

Potable reuse refers to reused water you can drink.

Nonpotable reuse refers to reused water that is not used for drinking, but is safe to use for irrigation or industrial purposes.

De facto, unacknowledged or unplanned potable reuse occurs when water intakes draw raw water supplies downstream from discharges of treated effluent from wastewater treatment plants/water reclamation facilities or resource recovery facilities. For example, if you are downstream of a community, that community's used water (run-off and treated wastewater) gets put back into river or stream and is delivered downstream to your community and becomes part of your drinking water supply.

Planned potable reuse is publicly acknowledged as an intentional project to recycle water for drinking water. It can be either direct or indirect. It commonly involves a more formal public process and public consultation program than is observed with de facto or unacknowledged reuse.

How potable reused water is delivered determines if it is called Indirect Potable Reuse or Direct Potable Reuse.

- Indirect potable reuse means the water is delivered to you indirectly. After it is purified, the reused water blends with other supplies and/or sits a while in some sort of storage, man-made or natural, before it gets delivered to a pipeline that leads to a water treatment plant or distribution system. That storage could be a groundwater basin or a surface water reservoir.
- Direct potable reuse means the reused water is put directly into pipelines that go to a water treatment plant or distribution system. Direct potable reuse may occur with or without "engineered storage" such as underground or above ground tanks.

Greywater uses the same waste as water reclamation with the exception of toilet water. Greywater can be used for irrigation, toilet flushing or other domestic uses.

Desalination is an energy-intensive process where salt and other minerals are removed from sea water to produce potable water for drinking and irrigation, typically through membrane filtration (reverse-osmosis), and steam-distillation.

Maximum water recovery - To determine maximum water recovery there are various techniques that have been developed by researchers; for maximum water reuse/reclamation/recovery strategies such as water pinch analysis. The techniques help a user to target the minimum freshwater consumption and wastewater target. It also helps in designing the network that achieves the target. This provides a benchmark to be used by users in improving their water systems.

Applications

Cover for reclaimed water valve, San Francisco Water District.

Most of the uses of water reclamation are non potable uses such as: Washing Cars, flushing toilets, cooling water for power plants, concrete mixing, artificial lakes, irrigation for golf courses and public parks, and for hydraulic fracturing. Where applicable, systems run a dual piping system to keep the recycled water separate from the potable water.

New technologies for recycling allow the water to be used for fracking purposes and can save an estimated 4 - 7 million or more gallons per well.

Potable Uses

Some water agencies reuse highly treated effluent from municipal wastewater or resource recovery plants as a reliable, drought proof source of drinking water. By using advanced purification processes, they produce water that meets all applicable drinking water standards. System reliability and frequent monitoring and testing are imperative to them meeting stringent controls.

The water needs of a community, water sources, public health regulations, costs, and the types of water infrastructure in place, such as distribution systems, man-made reservoirs, or natural groundwater basins, determine if and how reclaimed water can be part of the drinking water supply. Communities in El Paso, Texas and Orange County, California, for example, reuse water to replenish groundwater basins. Others, such as the Upper Occoquan Service Authority in Virginia, put it into surface water reservoirs. In these instances the reclaimed water is blended with other water supplies and/or sits in storage for a certain amount of time before it is drawn out and gets treated again at a water treatment or distribution system. In some Texas communities, the reused water is put directly into pipelines that go to a water treatment plant or distribution system. In Singapore reclaimed water is called NEWater and is bottled directly from an advanced water purification facility for educational and celebratory purposes. Though most of the reused water is used for high-tech industry in Singapore, a small amount is returned to reservoirs for drinking water.

A 2012 study conducted by the National Research Council in the United States of America found that the risk of exposure to certain microbial and chemical contaminants from drinking reclaimed water does not appear to be any higher than the risk experienced in at least some current drinking water treatment systems, and may be orders of magnitude lower. This report recommends adjustments to the federal regulatory framework that could enhance public health protection for both planned and unplanned (or de facto) reuse and increase public confidence in water reuse.

Modern technologies such as reverse osmosis and ultraviolet disinfection are commonly used when reclaimed water will be mixed with the drinking water supply. An experiment by the University of New South Wales reportedly showed a reverse osmosis system removed ethinylestradiol and paracetamol from the wastewater, even at 1000 times the expected concentration.

Aboard the International Space Station, astronauts have been able to drink recycled urine due to the introduction of the ECLSS system. The system costs $250 million and has been working since May 2009. The system recycles wastewater and urine back into potable water used for drinking, food preparation, and oxygen generation. This cuts back on the need for resupplying the space station so often.

Indirect Potable Reuse

Some municipalities are using and others are investigating Indirect Potable Reuse (IPR) of reclaimed water. For example, reclaimed water may be pumped into (subsurface recharge) or percolated down to (surface recharge) groundwater aquifers, pumped out, treated again, and finally used as drinking water. This technique may also be referred to as *groundwater recharging*. This includes slow processes of further multiple purification steps via the layers of earth/sand (absorption) and microflora in the soil (biodegradation).

Direct Potable Reuse

In a Direct Potable Reuse (DPR) scheme, water is put directly into pipelines that go to a water treatment plant or distribution system. Direct potable reuse may occur with or without "engineered storage" such as underground or above ground tanks. Communities in Texas have implemented DPR projects, and the state of California is studying the feasibility of developing DPR regulations.

Unplanned Potable Reuse

Water reuse occurs in various ways throughout the world. It happens daily on rivers and other water bodies everywhere. If you live in a community downstream of another, chances are you are reusing its water and likewise communities downstream of you are most likely reusing your water. Unplanned Indirect Potable Use has existed even before the introduction of reclaimed water. Many cities already use water from rivers that contain effluent discharged from upstream sewage treatment plants. There are many large towns on the River Thames upstream of London (Oxford, Reading, Swindon, Bracknell) that discharge their treated sewage ("non-potable water") into the river, which is used to supply London with water downstream.

This phenomenon is also observed in the United States, where the Mississippi River serves as both the destination of sewage treatment plant effluent and the source of potable water. Research conducted in the 1960s by the London Metropolitan Water Board demonstrated that the maximum extent of recycling water is about 11 times before the taste of water induces nausea in sensitive individuals. This is caused by the buildup of inorganic ions such as Cl^-, $SO4^{2-}$, K^+ and Na^+, which are not removed by conventional sewage treatment.

Space Travel

Wastewater reclamation can be especially important in relation to human spaceflight.

- In 1998, NASA announced it had built a human waste reclamation bioreactor designed for use in the International Space Station and a manned Mars mission. Human urine and feces are input into one end of the reactor and pure oxygen, pure water, and compost (humanure) are output from the other end. The soil could be used for growing vegetables, and the bioreactor also produces electricity.

Design Considerations

Distribution

Nonpotable reclaimed water is often distributed with a dual piping network that keeps reclaimed water pipes completely separate from potable water pipes. In the United States and some other countries, nonpotable reclaimed water is distributed in lavender (light purple) pipes to distinguish it from potable water. The use of the color purple for pipes carrying recycled water was pioneered by the Irvine Ranch Water District in Irvine, California.

In many cities using reclaimed water, it is now in such demand that consumers are only allowed to use it on assigned days. Some cities that previously offered unlimited reclaimed water at a flat rate are now beginning to charge citizens by the amount they use.

Reclamation Processes

Wastewater must pass through numerous systems before being returned to the environment. Here is a partial listing from one particular plant system:

- Barscreens - Barscreens remove large solids that are sent into a grinder. All solids are then dumped into a sewer pipe at a Treatment Plant.

- Primary Settling Tanks - Readily settable and floatable solids are removed from the wastewater. These solids are skimmed from the top and bottom of the tanks and sent to the Treatment Plant where it'll be turned into fertilizer.
- Biological Treatment - The wastewater is cleaned through a biological treatment method that uses microorganisms, bacteria which digest the sludge and reduce the nutrient content. Air bubbles up to keep the organisms suspended and to supply oxygen to the aerobic bacteria so they can metabolize the food, convert it to energy, CO_2, and water, and reproduce more microorganisms. This helps to remove ammonia also through nitrification.
- Secondary Settling Tanks - The force of the flow slows down as sewage enters these tanks, allowing the microorganisms to settle to the bottom. As they settle, other small particles suspended in the water are picked up, leaving behind clear wastewater. Some of the microorganisms that settle to the bottom are returned to the system to be used again.
- Tertiary Treatment - Deep-bed, single-media, gravity sand filters receive water from the secondary basins and filter out the remaining solids. As this is the final process to remove solids, the water in these filters is almost completely clear.
- Chlorine Contact Tanks - Three chlorine contact tanks disinfect the water to decrease the risks associated with discharging wastewater containing human pathogens. This step protects the quality of the waters that receive the wastewater discharge.

One of two procedures are then followed according to the future disposal site:

1. Reclaimed Water Pump Station - The pump station distributes reclaimed water to users around the City. This may include golf courses, agricultural uses, cooling towers, or in land fills.
2. Water is passed through high level purification to be returned to the environment. Currently this means a reverse osmosis system.

Treatment Improvements

As world populations require both more clean water and better ways to dispose of wastewater, the demand for water reclamation will increase. Future success in water reuse will depend on whether this can be done without adverse effects on human health and the environment.

In the United States, reclaimed waste water is generally treated to secondary level when used for irrigation, but there are questions about the adequacy of that treatment. Some leading scientists in the main water society, AWWA, have long believed that secondary treatment is insufficient to protect people against pathogens, and recommend adding at least membrane filtration, reverse osmosis, ozonation, or other advanced treatments for irrigation water.

There have been recent advances in reverse osmosis in different countries, but have consistently produced very high quality water all the same. In Singapore, reclaimed water, also known as NE-Water has become cleaner than the government issue tap water. Also, according to Bartels, the

Bedok Demonstration Plant, which uses RO membranes, has successfully run for the past 3 years, producing high quality wastewater all the while.

Seepage of nitrogen and phosphorus into ground and surface water is also becoming a serious problem, and will probably lead to at least tertiary treatment of reclaimed water to remove nutrients in the future. Even using secondary treatment, water quality can be improved. Water quality can also be improved as it passes through the subsurface mixing zone where surface water and groundwater combine. Testing for pathogens using Polymerase Chain Reaction (PCR) instead of older culturing techniques, and changing the discredited fecal coloform "indicator organism" standard would be improvements.

In a large study treatment plants showed that they could significantly reduce the numbers of parasites in effluent, just by making adjustments to the currently used process. But, even using the best of current technology, risk of spreading drug resistance in the environment through wastewater effluent, would remain.

Some scientists have suggested that there need to be basic changes in treatment, such as using bacteria to degrade waste based on nitrogen (urine) and not just carbonaceous (fecal) waste, saying that this would greatly improve effectiveness of treatment. Currently designed plants do not deal well with contaminants in solution (e.g. pharmaceuticals). "Dewatering" solids is a major problem. Some wastes could be disposed of without mixing them with water to begin with. In an interesting innovation, solids (sludge) could be removed before entering digesters and burned into a gas that could be used to run engines.

Emerging disinfection technologies include ultrasound, pulse arc electrohydraulic discharge, and bank filtration. Another issue is concern about weakened mandates for pretreatment of industrial wastes before they are made part of the municipal waste stream. Some also believe that hospitals should treat their own wastes. The safety of drinking reclaimed water which has been given advanced treatment and blended with other waters, remains controversial.

In recent years, as hydraulic fracturing of oil and gas formations has become more and more common place, new technologies for water recycling have emerged. One such technology uses a combination of ozone and electrocoagulation. This process removes organics, hydrocarbons, spent polymers, chemical additives used in the fracturing process, and heavy metals such as barium, iron, boron and more.

Alternatives

Seawater Desalination

In urban areas where climate change has threatened long-term water security and reduced rainfall over catchment areas, using reclaimed water for indirect potable use may be superior to other water supply augmentation methods. One other commonly used option is seawater desalination. Recycling wastewater and desalinating seawater may have many of the same disadvantages, including high costs of water treatment, infrastructure construction, transportation, and waste disposal problems. Although the best option varies from region to region, desalination is often superior economically, as reclaimed water usually requires a dual piping network, often with additional storage tanks, when used for nonpotable use.

Greywater Systems

A less elaborate alternative to reclaimed water is a greywater system. Greywater is wastewater that has been used in sinks, baths, showers, or washing machines, but does not contain sewage and has not been treated at the same levels as recycled water. In a home system, treated or untreated greywater may be used to flush toilets or for irrigation. Some systems now exist which directly use greywater from a sink to flush a toilet.

Rainwater Harvesting

Perhaps the simplest option is a rainwater harvesting system. Although there are concerns about the quality of rainwater in urban areas, due to air pollution and acid rain, many systems exist now to use untreated rainwater for nonpotable uses or treated rainwater for direct potable use. Urban design systems which incorporate rainwater harvesting and reduce runoff are known as Water Sensitive Urban Design (WSUD) in Australia, Low Impact Development (LID) in the United States and Sustainable urban drainage systems (SUDS) in the United Kingdom. There are also concerns about rainwater harvesting systems reducing the amount of run-off entering natural bodies of water.

Health Aspects

Reclaimed water is highly engineered for safety and reliability so that the quality of reclaimed water is more predictable than many existing surface and groundwater sources. Reclaimed water is considered safe when appropriately used. Reclaimed water planned for use in recharging aquifers or augmenting surface water receives adequate and reliable treatment before mixing with naturally occurring water and undergoing natural restoration processes. Some of this water eventually becomes part of drinking water supplies.

A water quality study published in 2009 compared the water quality differences of reclaimed/recycled water, surface water, and groundwater. Results indicate that reclaimed water, surface water, and groundwater are more similar than dissimilar with regard to constituents. The researchers tested for 244 representative constituents typically found in water. When detected, most constituents were in the parts per billion and parts per trillion range. DEET (a bug repellant), and Caffeine were found in all water types and virtually in all samples. Triclosan (in anti-bacterial soap & toothpaste) was found in all water types, but detected in higher levels (parts per trillion) in reclaimed water than in surface or groundwater. Very few hormones/steroids were detected in samples, and when detected were at very low levels. Haloacetic acids (a disinfection by-product) were found in all types of samples, even groundwater. The largest difference between reclaimed water and the other waters appears to be that reclaimed water has been disinfected and thus has disinfection by-products (due to chlorine use).

A 2005 study titled “Irrigation of Parks, Playgrounds, and Schoolyards with Reclaimed Water” found that there had been no incidences of illness or disease from either microbial pathogens or chemicals, and the risks of using reclaimed water for irrigation are not measurably different from irrigation using potable water. Studies by the National Academies of Science, the Monterey Regional Water Pollution Control Agency, and others have found reclaimed water to be safe for agricultural use.

Testing Standards

Reclaimed water is not regulated by the Environmental Protection Agency (EPA), but the EPA has developed water reuse guidelines that were most recently updated in 2012. The EPA Guidelines for Water Reuse represents the international standard for best practices in water reuse. The document was developed under a Cooperative Research and Development Agreement between the U.S. Environmental Protection Agency (EPA), the U.S. Agency for International Development (USAID), and the global consultancy CDM Smith. The Guidelines provide a framework for states to develop regulations that incorporate the best practices and address local requirements.

Ongoing wastewater research sometimes raise concerns about pathogens in the water. Many pathogens cannot be detected by currently used tests.

Recent literature also questions the validity of testing for "indicator organisms" instead of pathogens. Nor do present standards consider interactions of heavy metals and pharmaceuticals which may foster the development of drug resistant pathogens in waters derived from sewage.

To address these concerns about the source water, reclaimed water providers use multi-barrier treatment processes and constant monitoring to ensure that reclaimed water is safe and treated properly for the intended end use.

Potable Use

The main health risk for potable use of reclaimed water is the potential for pharmaceutical and other household chemicals or their derivatives (Environmental persistent pharmaceutical pollutants) to persist in this water. This would be of much less concern if the population were to keep their excrement out of the wastewater e.g. via the use of the Urine-diverting dry toilet or systems that treat blackwater separately from greywater.

Environmental Aspects

There is debate about possible health and environmental effects. To address these concerns, A Risk Assessment Study of potential health risks of recycled water and comparisons to conventional Pharmaceuticals and Personal Care Product (PPCP) exposures was conducted by the WateReuse Research Foundation. For each of four scenarios in which people come into contact with recycled water used for irrigation - children on the playground, golfers, and landscape, and agricultural workers - the findings from the study indicate that it could take anywhere from a few years to millions of years of exposure to nonpotable recycled water to reach the same exposure to PPCPs that we get in a single day through routine activities.

Using reclaimed water for non-potable uses saves potable water for drinking, since less potable water will be used for non-potable uses.

It sometimes contains higher levels of nutrients such as nitrogen, phosphorus and oxygen which may somewhat help fertilize garden and agricultural plants when used for irrigation.

The usage of water reclamation decreases the pollution sent to sensitive environments. It can also enhance wetlands, which benefits the wildlife depending on that eco-system. It also helps to stop the chances of drought as recycling of water reduces the use of fresh water supply from underground sources. For instance, The San Jose/Santa Clara Water Pollution Control Plant instituted a water recycling program to protect the San Francisco Bay area's natural salt water marshes.

Costs and Evaluation

The cost of reclaimed water exceeds that of potable water in many regions of the world, where a fresh water supply is plentiful. However, reclaimed water is usually sold to citizens at a cheaper rate to encourage its use. As fresh water supplies become limited from distribution costs, increased population demands, or climate change reducing sources, the cost ratios will evolve also. The evaluation of reclaimed water needs to consider the entire water supply system, as it may bring important value of flexibility into the overall system.

History

Storm and sanitary sewers were necessarily developed along with the growth of cities. By the 1840s the luxury of indoor plumbing, which mixes human waste with water and flushes it away, eliminated the need for cesspools. Odor was considered the big problem in waste disposal and to address it, sewage could be drained to a lagoon, or "settled" and the solids removed, to be disposed of separately. This process is now called "primary treatment" and the settled solids are called "sludge."

At the end of the 19th century, since primary treatment still left odor problems, it was discovered that bad odors could be prevented by introducing oxygen into the decomposing sewage. This was the beginning of the biological aerobic and anaerobic treatments which are fundamental to waste water processes.

By the 1920s, it became necessary to further control the pollution caused by the large quantities of human and industrial liquid wastes which were being piped into rivers and oceans, and modern treatment plants were being built in the US and other industrialized nations by the 1930s.

Designed to make water safe for fishing and recreation, the Clean Water Act of 1972 mandated elimination of the discharge of untreated waste from municipal and industrial sources, and the US federal government provided billions of dollars in grants for building sewage treatment plants around the country. Modern treatment plants, usually using oxidation and/or chlorination in addition to primary and secondary treatment, were required to meet certain standards.

Current treatment improves the quality of separated wastewater solids or sludge. The separated water is given further treatment considered adequate for non potable use by local agencies, and discharged into bodies of water, or reused as reclaimed water. In places like Florida, where it is necessary to avoid nutrient overload of sensitive receiving water, reuse of treated or reclaimed water can be more economically feasible than meeting the higher standards for surface water disposal mandated by the Clean Water Act.

Examples

Indirect Potable Reuse (IPR)

- Orange County, California.

- Pasadena, California.
- Singapore (where it is branded as *NEWater*).
- Payson, Arizona.
- The Torreele project in the Veurne coastal region of Belgium, which began operating in 2002.
- Virginia Occoquan Reservoir - The Upper Occoquan Sewage Authority plant discharges its highly treated output to supply roughly 20% of the inflow into the Occoquan Reservoir, which provides drinking water used by the Fairfax County Water Authority - one of the three major water providers in the Washington, D.C. metropolitan area.

Non-potable Reuse (NPR)

- Austin, Texas.
- Caboolture and Maroochy (South East Queensland, Australia) LGA's currently provide reclaimed water for industrial use (primarily capital works). Users must apply for a key to be able to access the compounds in which the outlets are located.
- Clark County, Nevada.
- Clearwater, Florida.
- Contra Costa County, California.
- Melbourne, Australia.
- Mount Buller Ski resort uses recycled water for snow making.
- San Antonio operates the largest recycled water system in the United States.
- Sydney, Australia.
- Tucson, Arizona.
- San Diego, California (San Diego County).
- St. Petersburg, Florida.

Proposed

In some places, reclaimed water has been proposed for either potable or non-potable use:

- South East Queensland, Australia (planned for potable use as of late 2010).
- Newcastle, New South Wales, Australia (proposed for non-potable use).
- Canberra, Australian Capital Territory, Australia (proposed in January 2007 as a backup source of potable water).

- Los Angeles, California - By 2019, the Los Angeles Department of Water and Power will build a plant to replenish their groundwater aquifer with purified water in order to deal with the shortage of rain and snow fall, restricted water imports and local groundwater contamination.

Israel

As of 2010, Israel leads the world in the proportion of water it recycles. Israel treats 80% of its sewage (400 billion liters a year), and 100% of the sewage from the Tel Aviv metropolitan area is treated and reused as irrigation water for agriculture and public works. The remaining sludge is currently pumped into the Mediterranean, however a new bill has passed stating a conversion to treating the sludge to be used as manure. Only 20% of the treated water is lost (due to evaporation, leaks, overflows and seeping). The recycled water allows farmers to plan ahead and not be limited by water shortages. There are many levels of treatment, and many different ways of treating the water—which leads to a big difference in the quality of the end product. The best quality of reclaimed sewage water comes from adding a gravitational filtering step, after the chemical and biological cleansing. This method uses small ponds in which the water seeps through the sand into the aquifer in about 400 days, then is pumped out as clear purified water. This is nearly the same process used in the space station water recycling system, which turns urine and feces into purified drinking water, oxygen and manure.

To add to the efficiency of the Israeli system - the reclaimed sewage water may be mixed with reclaimed sea water (Plans are in action to increase the desalinization program up to 50% of the countries usage by 2013 - 600 billion liters of drinkable sea water a year), along with aquifer water and fresh sweet lake water - monitored by computer to account for the nationwide needs and input. This action reduced the outdated risk of salt and mineral percentages in the water. Plans to implement this overall usage of reclaimed water for drinking are discouraged by the psychological preconception of the public for the quality of reclaimed water, and the fear of its origin. As of today, all the reclaimed sewage water in Israel is used for agricultural and land improvement purposes.

U.S.

The leaders in use of reclaimed water in the U.S. are Florida and California, with Irvine Ranch Water District as one of the leading developers. They were the first district to approve the use of reclaimed water for in-building piping and use in flushing toilets.

In a January 2012 U.S. National Research Council report, a committee of independent experts found that expanding the reuse of municipal wastewater for irrigation, industrial uses, and drinking water augmentation could significantly increase the United States' total available water resources. The committee noted that a portfolio of treatment options is available to mitigate water quality issues in reclaimed water. The report also includes a risk analysis that suggests the risk of exposure to certain microbial and chemical contaminants from drinking reclaimed water is not any higher than the risk from drinking water from current water treatment systems—and in some cases, may be orders of magnitude lower. The report concludes that adjustments to the federal regulatory framework could enhance public health protection and increase public confidence in water reuse.

Australia

As Australia continues to battle the 7–10-year drought, nationwide, reclaimed effluent is becoming a popular option. Two major capital cities in Australia, Adelaide and Brisbane, have already committed to adding reclaimed effluent to their dwindling dams. The former has also built a desalination plant to help battle any future water shortages. Brisbane has been seen as a leader in this trend, and other cities and towns will review the Western Corridor Recycled Water Project once completed. Goulbourn, Canberra, Newcastle, and Regional Victoria, Australia are already considering building a reclaimed effluent process.

European Union

The second largest waste reclamation program in the world is in Spain, where 12% of the nation's waste is treated.

According to an EU-funded study, "Europe and the Mediterranean countries are lagging behind" California, Japan, and Australia "in the extent to which reuse is being taken up." According to the study "the concept (of reuse) is difficult for the regulators and wider public to understand and accept."

Farm Water

Farm water, also known as agricultural water, is water committed for use in the production of food and fiber. On average, 80 percent of the fresh water withdrawn from rivers and groundwater is used to produce food and other agricultural products. Farm water may include water used in the irrigation of crops or the watering of livestock.

It becomes difficult to relate water use to economics, however, water is one of the most fundamental parts of the global economy. It has been shown, that in areas without healthy water resources or sanitation services that economic growth cannot be sustained. Without access to clean water, nearly every industry would suffer, most notably, would be agriculture. As water scarcity grows as a global concern, food security is also brought into consideration. A recent example of this could be the drought in California—it is projected that for every $100 spent on foods from this state, a consumer should expect to pay up to $15 additionally.

Livestock Water Use

Livestock and meat production have one of the largest water footprints of the agricultural industry. Taking nearly 1,800 gallons of water to produce one pound of beef and 576 gallons for pork. For reference, it only takes about 108 gallons of water to harvest one pound of corn. Livestock production is also one of the most resource intensive agricultural outputs. This is largely due to their large feed conversion ratio. Livestock's large water consumption may also be attributed to the amount of time it takes to raise an animal to slaughter. Again, in contrast to corn, which grows to maturity in about 100 days. This adds an extra 995 days of water to grow cattle. The global "food animal" population is just over 20 billion creatures, with 7+ billion humans, this equates to about 2.85 animals per human.

Cattle

The beef and dairy industries are the most lucrative branches of the U.S. agricultural industry, however, they are also the most resource intensive. To date, beef is the most popular of the meats; the U.S. alone produced 25.8 billion pounds in 2013. In this same year, 201.2 billion pounds of milk were produced. These cattle are mostly raised in centralized animal feeding operations, or CAFOs. On average, a mature cow will consume anywhere from seven to twenty-four gallons of water a day; cows that are lactating require about twice as much water. The amount of water that cattle may drink in a day also depends upon the temperature. Cattle have a feed conversion ratio of 6:1, for every six pounds of food consumed, the animal should gain one pound. Thus, there is also a substantial "indirect" need for water in order to grow the feed for the livestock. Growing the amount of feed grains necessary for raising livestock accounts for 56 percent of the U.S water consumption. Of a 1,000 pound cow, only 430 pounds make it to the retail markets. This 18 percent loss, creates an even greater demand for cattle, being that CAFOs must make up for this lost profitable weight, by increasing the amount of cows that they raise.

Water scarcity is not necessarily a new issue, however, cattle ranchers in America have been cutting heard sizes since the 1950s in efforts to curb water and manufacturing costs. This shift has led to more efficient feeding and health methods, allowing ranchers to harvest more beef per animal. The rising popularity of these CAFOs are creating a larger demand for water, however. Grass-fed or grazing cows consume about twelve percent more water through the ingestion of live plants, than those cows who are fed dried grains.

Poultry and Fowl

Water is one of the most crucial aspects of poultry raising, as they use this to carry food through their system, and assist in digestion. Additionally, farmers monitor flock water consumption to measure the overall health of their birds. As birds grow older they should be consuming more feed and about three times as much water. In just three weeks a bird's water consumption should increase by about 10 gallons a day. Water consumption is also influenced by temperature. When it is hot outside bird will pant to keep cool, thus losing much of their water. A study based in Ohio showed that 67 percent of water sampled near poultry farms contained antibiotics.

Horticulture Water Use

With modern advancements, crops are being cultivated year round in countries all around the world. As water usage becomes a more pervasive global issue, irrigation practices for crops are being refined and becoming more sustainable. While there are a variety of irrigation systems, these may be grouped into two types: high flow and low flow. These systems must be managed precisely to prevent runoff, over spray, or low-head drainage.

Scarcity of Water in Agriculture

50 years ago, the common perception was that water was an infinite resource. At this time, there were fewer than half the current number of people on the planet. Affluence was not as high, individuals consumed fewer calories and ate less meat, so less water was needed to produce their food. They required a third of the volume of water we presently take from rivers. Today, the competition for water resources is much more intense. This is because there are now nearly eight billion people on the planet, their consumption of meat and vegetables is rising. Competition for water from industry, urbanisation and biofuel crops is rising congruently. To avoid a global water crisis, farmers will have to make strides to increase productivity to meet growing demands for food, while industry and cities find ways to use water more efficiently.

Successful agriculture is dependent upon farmers having sufficient access to water. However, water scarcity is already a critical constraint to farming in many parts of the world. Physical water scarcity is where there is not enough water to meet all demands, including that needed for ecosystems to function effectively. Arid regions frequently suffer from physical water scarcity. It also occurs where water seems abundant but where resources are over-committed. This can happen where there is over development of hydraulic infrastructure, usually for irrigation. Symptoms of physical water scarcity include environmental degradation and declining groundwater. Economic scarcity, meanwhile, is caused by a lack of investment in water or insufficient human capacity to satisfy the demand for water. Symptoms of economic water scarcity include a lack of infrastructure, with people often having to fetch water from rivers for domestic and agricultural uses. Some 2.8 billion people currently live in water-scarce areas.

Sustainable Water Use

While it is evident that water use affects environmental degradation and economic growth, it is also sparking innovation regarding new irrigation methods. In 2006, the USDA predicted that if the agricultural sector improved water efficiency by just 10 percent farms could annually save upwards of $200 million per year. Many of the practices which cut water use are cost effective. Farmers who use straw, compost, or mulch around their crops will reduce evaporation by about 75 percent. This would also reduce the number of weeds and save a farmer from using herbicides. Mulches or ground covers also allow the soils to absorb more water by reducing compaction. The use of white or pale gravel is also practiced, as it reduces evaporation and keeps soil temperatures low by reflecting sunlight.

In addition to reducing water loss at the sink, there are also more sustainable ways to harvest water. Many modern small (non industrial) farmers are using rain barrels to collect the water needed for their crops and livestock. On average, rainwater harvesting reduces the cost of water per cubic meter in half. This would also greatly reduce the stress on local aquifers and wells. Because farmers use the roofs of their buildings to gather this water, this also reduced rainwater runoff and soil erosion on and around their farm.

References

- Water Reuse: Potential for Expanding the Nation's Water Supply through Reuse of Municipal Wastewater.

National Research Council. 2012. ISBN 978-0-309-25749-7

- "Prevent weeds, improve soil and save water by mulching your garden | Oregon State University Extension Service | Gardening". extension.oregonstate.edu. Retrieved 02, August 2020
- Post, V. E. A.; Groen, J.; Kooi, H.; Person, M.; Ge, S.; Edmunds, W. M. (2013). "Offshore fresh groundwate reserves as a global phenomenon". Nature. 504 (7478): 71–78. doi:10.1038/nature12858. PMID 24305150
- "Water Recycling and Reuse: The Environmental Benefits/". US Environment Protection Agency. 23 February 2016. Retrieved 22, August 2020
- Helgeson, Tom (2009). A Reconnaissance-Level Quantitative Comparison of Reclaimed Water, Surface Water, and Groundwater. Alexandria, VA: WateReuse Research Foundation. p. 141. ISBN 978-1-934183-12-0
- "Urban Trees Enhance Water Infiltration". Fisher, Madeline. The American Society of Agronomy. November 17, 2008. Retrieved 31, January 2012
- Crook, James (2005). Irrigation of Parks, Playgrounds, and Schoolyards: Extent and Safety. Alexandria, VAr WateReuse Research Foundation. p. 60. ISBN 0-9747586-3-9
- "What is hydrology and what do hydrologists do?". The USGS Water Science School. United States Geological Survey. 23 May 2013. Retrieved 21 January 2020

Permissions

All chapters in this book are published with permission under the Creative Commons Attribution Share Alike License or equivalent. Every chapter published in this book has been scrutinized by our experts. Their significance has been extensively debated. The topics covered herein carry significant information for a comprehensive understanding. They may even be implemented as practical applications or may be referred to as a beginning point for further studies.

We would like to thank the editorial team for lending their expertise to make the book truly unique. They have played a crucial role in the development of this book. Without their invaluable contributions this book wouldn't have been possible. They have made vital efforts to compile up to date information on the varied aspects of this subject to make this book a valuable addition to the collection of many professionals and students.

This book was conceptualized with the vision of imparting up-to-date and integrated information in this field. To ensure the same, a matchless editorial board was set up. Every individual on the board went through rigorous rounds of assessment to prove their worth. After which they invested a large part of their time researching and compiling the most relevant data for our readers.

The editorial board has been involved in producing this book since its inception. They have spent rigorous hours researching and exploring the diverse topics which have resulted in the successful publishing of this book. They have passed on their knowledge of decades through this book. To expedite this challenging task, the publisher supported the team at every step. A small team of assistant editors was also appointed to further simplify the editing procedure and attain best results for the readers.

Apart from the editorial board, the designing team has also invested a significant amount of their time in understanding the subject and creating the most relevant covers. They scrutinized every image to scout for the most suitable representation of the subject and create an appropriate cover for the book.

The publishing team has been an ardent support to the editorial, designing and production team. Their endless efforts to recruit the best for this project, has resulted in the accomplishment of this book. They are a veteran in the field of academics and their pool of knowledge is as vast as their experience in printing. Their expertise and guidance has proved useful at every step. Their uncompromising quality standards have made this book an exceptional effort. Their encouragement from time to time has been an inspiration for everyone.

The publisher and the editorial board hope that this book will prove to be a valuable piece of knowledge for students, practitioners and scholars across the globe.

Index

www.ingramcontent.com/pod-product-compliance
Lightning Source LLC
Chambersburg PA
CBHW081415220926
51613CB00010B/237
* 9 7 8 1 6 3 9 8 7 3 3 8 8 *